Media, Terrorism and Society

This book provides new insights on contemporary terrorism and media research, opening the door for fresh perspectives and trends exploring theories and concepts in the field. Advances in technology have increased the threat of terrorism, as the Internet has helped terrorists to recruit new members, plan their attacks, and amplify their messages. As technology continues to evolve, it is not difficult to imagine how the advanced information and technology of the new millennium could cause more terrifying realities in the world today. During this period of profound technological change, we need to understand the relationships between media, society, and the new paradigm of terrorism. In our global society where the war on terrorism knows no borders, countries are increasingly recognizing the importance of improving terrorism coverage domestically and globally. This book is a valuable resource, offering key directions for assessing the ongoing revolutionary changes and trends in communicating terrorism in the digital age.

This book was originally published as a special issue of *Mass Communication and Society*.

Shahira S. Fahmy is an internationally renowned scholar in peace journalism and visual communication. She is the first and only Arab-American journalism professor tenured at an American research university. Her seminal research has appeared in all the top-ranked journals, and one of her books received the National Communication Association book award.

Media, Terrorism and Society

Perspectives and Trends in the Digital Age

Edited by
Shahira S. Fahmy

Routledge
Taylor & Francis Group
LONDON AND NEW YORK

First published 2019
by Routledge
2 Park Square, Milton Park, Abingdon, Oxon, OX14 4RN, UK

and by Routledge
52 Vanderbilt Avenue, New York, NY 10017, USA

Routledge is an imprint of the Taylor & Francis Group, an informa business

British Library Cataloguing in Publication Data
A catalogue record for this book is available from the British Library

ISBN 13: 978-1-138-36064-8

Typeset in Times New Roman
by RefineCatch Limited, Bungay, Suffolk

Publisher's Note
The publisher accepts responsibility for any inconsistencies that may have
arisen during the conversion of this book from journal articles to book chapters,
namely the possible inclusion of journal terminology.

Disclaimer
Every effort has been made to contact copyright holders for their permission to
reprint material in this book. The publishers would be grateful to hear from any
copyright holder who is not here acknowledged and will undertake to rectify
any errors or omissions in future editions of this book.

MIX
Paper from
responsible sources
FSC™ C013985
www.fsc.org

Printed in the United Kingdom
by Henry Ling Limited

Contents

CONTENTS

Citation Information

The chapters in this book were originally published in *Mass Communication and Society*, volume 20, issue 6 (November–December 2017). When citing this material, please use the original page numbering for each article, as follows:

Guest Editor's Note

Media, Terrorism, and Society: Perspectives and Trends in the Digital Age
Shahira S. Fahmy
Mass Communication and Society, volume 20, issue 6 (November–December 2017), pp. 735–739

Chapter 1

Solidarity Through the Visual: Healing Images in the Brussels Terrorism Attacks
Dan Berkowitz
Mass Communication and Society, volume 20, issue 6 (November–December 2017), pp. 740–762

Chapter 2

Graphic Violence as Moral Motivator: The Effects of Graphically Violent Content in News
Matthew Grizzard, Jialing Huang, Julia K. Weiss, Eric Robert Novotny, Kaitlin S. Fitzgerald, Changhyun Ahn, Zed Ngoh, Alexandra Plante, and Haoran Chu
Mass Communication and Society, volume 20, issue 6 (November–December 2017), pp. 763–783

Chapter 3

Online Surveillance's Effect on Support for Other Extraordinary Measures to Prevent Terrorism
Elizabeth Stoycheff, Kunto A. Wibowo, Juan Liu, and Kai Xu
Mass Communication and Society, volume 20, issue 6 (November–December 2017), pp. 784–799

Chapter 4

The Impact of Terrorist Attack News on Moral Intuitions and Outgroup Prejudice
Ron Tamborini, Matthias Hofer, Sujay Prabhu, Clare Grall,
Eric Robert Novotny, Lindsay Hahn, and Brian Klebig
Mass Communication and Society, volume 20, issue 6 (November–December 2017), pp. 800–824

Chapter 5

"Muslims are not Terrorists": Islamic State Coverage, Journalistic Differentiation Between Terrorism and Islam, Fear Reactions, and Attitudes Toward Muslims
Christian von Sikorski, Desirée Schmuck, Jörg Matthes, and
Alice Binder
Mass Communication and Society, volume 20, issue 6 (November–December 2017), pp. 825–848

Chapter 6

On the Boundaries of Framing Terrorism: Guilt, Victimization, and the 2016 Orlando Shooting
Nathan Walter, Thomas J. Billard, and Sheila T. Murphy
Mass Communication and Society, volume 20, issue 6 (November–December 2017), pp. 849–868

Chapter 7

Proximity and Terrorism News in Social Media: A Construal-Level Theoretical Approach to Networked Framing of Terrorism in Twitter
K. Hazel Kwon, Monica Chadha, and Kirstin Pellizzaro
Mass Communication and Society, volume 20, issue 6 (November–December 2017), pp. 869–894

Chapter 8

U.S. News Coverage of Global Terrorist Incidents
Mingxiao Sui, Johanna Dunaway, David Sobek, Andrew Abad,
Lauren Goodman, and Paromita Saha
Mass Communication and Society, volume 20, issue 6 (November–December 2017), pp. 895–908

For any permission-related enquiries please visit:
http://www.tandfonline.com/page/help/permissions

Dedication

To the journalists who risk their lives to get their story

INTRODUCTION

Media, Terrorism, and Society: Perspectives and Trends in the Digital Age

Shahira S. Fahmy

With the advent of the new millennium, the number of terrorism attacks increased worldwide. The attacks inspired a series of empirical terrorism studies, and I received a wave of invitations to review the burgeoning research on media and terrorism. Certainly I could feel the topic was critically important, but I felt a troubling sense that behind the invitations was an assumption that my Middle Eastern background qualified me to pronounce on terrorism. And yet, unknown to most of the people asking me for help, they had tapped a raw nerve. During my university days I was the survivor of a terrorist attack, the witness of a deadly bombing when religious extremists detonated a bomb near Cairo's busiest square in Egypt.

Those haunting memories, coupled with a sense of global responsibility and academic enthusiasm, pushed me to ask Dr. Fuyuan Shen, the editor-in-chief of *Mass Communication and Society*, to offer a special issue on this topic. I felt this issue could offer a forum for inspiring new conversations and advancing studies pertaining to terrorism within the mass communication field. The proposal was competitively selected by Dr. Shen and his editorial board, and the response to the call was unprecedented—almost 50 national and international submissions solid, from which eight were selected after several rounds of peer reviews.

1

Whereas the literature offers no single embracing definition of terrorism, scholars can agree that terrorism includes the use of violence against civilians, with the intention of creating fear or terror, and the intention to force specific belief systems upon others. With the advance of technology, the threat of terrorism has been of growing concern, especially because the Internet has helped terrorists recruit new members, plan their attacks, and amplify their messages.

As all technology continues to evolve, it is not difficult to imagine how the advanced information and technology of the new millennium could cause more terrifying realities in the world today. Cooper (2001) explained that the onward sweep of technology makes it possible to undertake newer and more efficient ways to create the fear that is at the heart of terrorism.

Without doubt, during this period of profound technological changes, researchers should study the relationships among media, society, and the new paradigm of terrorism. As Cummings and Frost (1995) wrote, academic studies are an influencing source that legitimizes ideas and research findings. The studies here, therefore, provide valuable knowledge and offer key directions for current and future terrorism research within our discipline.

Research on media and terrorism is crucial, especially amid the increase in terrorism activities. According to the U.S. Department of State (National Consortium for the Study of Terrorism and Responses to Terrorism: Annex of Statistical Information, 2015), there has been a 35% increase in terrorist attacks across the world between 2013 and 2014, with the total number of people killed by terrorism increasing by more than 80%. Comparatively, however, investigating the role of media in communicating such trends has been a relatively new field of research. These special issue articles, which have all been selected for their unique contribution to the literature, display this fast-emerging research area.

In the call for papers, manuscripts were solicited that offered to build upon traditional approaches to mass media's role in shaping and amplifying terrorism issues by opening new space for including how the social and/or mass media have communicated terrorism, specifically with the evolution of digital media. Four of the eight studies look at the media effects of exposure to terrorism coverage that span domestic and global communities; three studies review roles and patterns in terrorism news. The remaining study looks at the relationship between online mass surveillance and terrorism prevention.

The special issue begins with Dan Berkowitz's work that applies a semiotic lens to examine the role images of terrorism play in mediating healing and solidarity in both the local and the global media arena. Using a qualitative examination of news reports and blogs related to the 2016 Brussels terrorist attacks, he found that a combination of photographs, editorial cartoons, and anchoring texts has the potential to build global solidarity against terrorism.

One of the insights of his work emphasizes the importance of the anchoring function. Berkowitz explains that this function is essential for audience decoding, specifically when photographs and cartoons are global, and therefore not readily accessible to the broader audience.

In the next article, Grizzard and colleagues take us a step further in examining the role of visuals with an empirical study that challenges common wisdom and assesses whether visuals containing graphic violence yield prosocial responses (e.g., greater moral sensitivity). The authors set the stage by conducting two experimental studies using news footage of a mass execution conducted by the Islamic State group. Noting that graphically violent visuals can have prosocial effects, they suggest that editorial policies governing the display of graphic violence should take a more balanced approach to the issue. It might be possible, they note, that only by confronting graphic visuals does one becomes motivated to put an end to the violence that caused it.

The work of Stoycheff, Wibowo, Liu, and Xu is a reminder of the threat that terrorism accentuates and how average Americans have continued to sacrifice online civil liberties for national security. Using an online experiment, the authors examine the relationship between perceptions of surveillance online and support for other extraordinary measures to prevent terrorism. Results suggest perceptions of government monitoring prompt support for restricting others' online and offline civil liberties, including rights to free speech and a fair trial. This support, in turn, contributes to a culture of mistrust and suspicion and feeds support for military intervention abroad in the name of terrorism prevention. The study contributes to the limited, but growing, body of empirical scholarship that reveals online mass surveillance produces spillover effects, including political intolerance and hawkish policy attitudes.

Tamborini and colleagues look at the impact of exposure to terrorist attack news coverage on the salience of moral intuitions and prosocial behavioral intentions toward outgroup members. They highlight the potential value of the model of intuitive morality exemplars for understanding the way media exposure can influence tolerance for other people and ideas. The authors report the impact of terrorist news exposure on increasing the salience of respect for authority (e.g., respect for and deference to traditions and hierarchies), which in turn decreases the willingness to help members of a group that one does not identify with.

Attitudes toward outgroup members comes in again for attention in a second empirical piece. Sikorski, Schmuck, Matthes, and Binder look at news coverage about Islamist terrorism and entertain questions about whether news differentiation (i.e., explicitly distinguishing between Muslims and Muslim terrorists) may dampen particular fear reactions. In line with the intergroup threat theory, results of a controlled laboratory experiment show that undifferentiated news about terrorism can come with serious consequences, namely, negative outgroup perceptions of Muslims in general. The authors suggest that these perceptions may

3

further enhance intergroup conflicts between Muslims and non-Muslims. One of the significant implications of their analyses is highlighting the importance of clearly and explicitly distinguishing between news about Muslims and Muslim terrorists. News differentiation, they argue, would result in a better informed and less hostile environment.

The effects of ingroup–outgroup dynamics are further explored by Walter, Billard, and Murphy. Building on a more holistic approach to the study of framing and terrorism, the authors anchor their study in media effects, collective memory, and social networks. They used coverage of the Orlando nightclub shooting to demonstrate how media frames (homophobic hate crime and Islamic terrorist attack) cement competing interpretations by evoking social categories and collective emotions. They report that increasing ingroup inclusiveness— from an attack on the lesbian, gay, bisexual, transgender, queer (LGBTQ) community to an attack on Americans—leads to greater collective victimization, subsequently undermining the need for reconciliation with the LGBT outgroup.

Kwon, Chadha, and Pellizzaro build on terrorism scholarship and social media research by examining the framing of terrorism news in Twitter. They look at the global nature of public's terrorism sense-making in the contemporary social media environment and delve into the role of news proximity in both audience frames and media frames of terrorism. An analysis of tweets during the Boston Marathon bombing and the Brussels Airport attack suggests that institutional and audience frames show a great deal of similarity but do not always converge on Twitter. With technology impacting global patterns of communication, such similarities may be attributed to a universal human tendency for social categorization and stereotyping, inherent not only in the minds of ordinary citizens but also in that of journalists. By including audience framing in the picture, the authors add a more complex understanding of people's cognitive response and subsequent information sharing on Twitter during a crisis, such as a terrorist attack.

Finally, the trends and patterns of U.S. news coverage of terrorist attacks are analyzed in the work of Sui and colleagues. Focusing on 15 years, from 1998 to 2013, they provide a unique longitudinal view. Noting the recent terrorism activities by the Islamic State group, the authors found that U.S. media are more likely to cover the group's non-U.S. events but fail to find similar results for other well-known terrorist organizations. Still, findings are consistent with expectations set forth by the literature on norms and routines of journalism and economics of news. The authors report that proximity to and affinity with the United States, weapons of mass destruction, and the number of global and U.S. casualties determine terrorism coverage by U.S. major media outlets.

Collectively, these studies provide new insights and open the discussions for fresh perspectives and trends exploring theories and concepts on contemporary terrorism and media research. Clearly, communication scholarship on terrorism adds to the field and is a significant body of literature in its own right. An objective of this special issue is to showcase the distinctive growth in terrorism studies and to spark additional research. Dealing with different topics, research questions, and new data, I hope this research will animate a new generation of scholars interested in one of the fastest growing interest areas in the communication discipline.

I thank Dr. Shen and his assistant editors, Pratiti Diddi and Lewen Wei, for their insights and support, as well as the authors who submitted manuscripts to this special issue. This work offers genuine new knowledge, and it would not have come to fruition without the creativity and hard work of its 30 contributors. My gratitude also goes to the 67 reviewers, who provided sound and constructive criticism to the authors, ranging from input on methodological analyses and design to theoretical reasoning.

While this special issue will not be the last one focusing on this topic, it is an essential one, nevertheless. In our global society where the war on terrorism knows no borders, countries are increasingly recognizing the importance of improving terrorism coverage domestically and globally, even as the traditional role of the media is becoming deemphasized. Associated Press Middle East Editor Dan Perry (personal communication, August 27, 2017) explained, "For a variety of reasons large segments of society are detaching from the professional media and are in the thrall of agenda media not beholden to journalistic standards and given to scandal and conspiracy." Our challenge, therefore, is to keep investigating the way emerging technologies are transforming the relationship between media and terrorism in society. This will always be an open-ended research endeavor, but our work should focus on reexamining existing theories while integrating new concepts that touch on this topic, because this research will help us understand and explain the ongoing revolutionary changes and trends in communicating terrorism.

REFERENCES

Cooper, H. H. A. (2001). Terrorism: The problem of definition revisited. *American Behavioral Scientist, 44*(6), 881–893.

Cummings, L. L., & Frost, P. J. (1995). *Publishing in the organizational sciences* (2nd ed., Foundations for Organizational Science Series). Thousand Oaks, CA: Sage.

National Consortium for the Study of Terrorism and Responses to Terrorism: Annex of Statistical Information. (2015). Bureau of Counterterrorism. *Country Reports on Terrorism 2014*. Retrieved from http://files.eric.ed.gov/fulltext/ED219724.pdf

Solidarity Through the Visual: Healing Images in the Brussels Terrorism Attacks

Dan Berkowitz

This study applies a semiotic lens to argue that news media are not only conveyors of images of terrorism but, through a combination of photographs, editorial cartoons, and anchoring texts, have the potential to serve as facilitators of healing and solidarity in both the local and the global media arena. This anchoring function is essential for audience decoding—especially in the global—because many photographs and cartoons are culturally bound and therefore not readily accessible to the broader audience. Through the information they provide, written texts help to anchor meanings of the visual texts, conveying a sense of solidarity against terrorism.

Terrorism—once an act confined to a relatively few world regions—has spread globally, and with it a wealth of visual images reflecting trauma, mourning, and pain (Kircher, 2016; Nacos, 2016). Photographs play a large role in conveying emotions and facilitating healing from grief (Kitch, 2000, 2002; Zelizer, 2002); so, too, do political cartoons (Denham, 2015). This was especially the case when three coordinated bombings took place at the Brussels, Belgium, airport and a Brussels metro station on March 22, 2016, causing 32 deaths and more than 300 injuries.

Dan Berkowitz (Ph.D., Indiana University, 1988) is a professor in the School of Journalism & Mass Communication at the University of Iowa. His research interests include social and cultural production of news, media and terrorism, news and collective memory, mythical news narratives, and journalistic boundary work.

Color versions of one or more of the figures in the article can be found online at www.tandfonline.com/hmcs.

6

This study argues that news media not only are conveyors of images of terrorism but also—through those images—have the potential to serve as mediators of healing and solidarity in the global arena (El Refaie, 2009; Fahmy, 2010; Hou, nd). However, news media face a challenge: Many photographs and cartoons are culturally bound and therefore not universally understood (Rose, 2016). The role of texts that accompany and discuss visual images thus play a crucial role in anchoring the meanings of images, helping to build understanding and a global sense of solidarity.

Data come from news reports and blogs from English-language media that addressed visual responses to the Brussels terrorist attacks. Among those visual elements were photos of memorial sites, political cartoons, and memes of the Belgian cartoon character Tintin and of an iconic fountain in Brussels. Analysis was guided by elements of semiotics and accomplished through a qualitative examination of news texts and the visual content itself.

CONCEPTUAL FOUNDATION

Research on communication about terrorism is a complex, cross-cultural topic. Nacos (2016) depicted terrorism communication as a relationship between terrorists and a three-way linkage between media, citizens, and government at both the national and international levels. In turn, these domestic and global linkages are all immersed within the Internet. As a result, terrorists have direct, easy, and inexpensive access to computer-aided communication. In the end, this web of communication serves terrorism with goals of both publicity and propaganda. Weimann (1983) called this "the theater of terror," where terrorists "write the script and perform the drama" that only becomes possible when media provide the stage and the audience (p. 38). This calculated action is often preplanned to provide conditions that media need in selecting what to cover. Through these acts of terrorism, terrorists create an environment of fear that helps leverage their position in a social or political conflict (Altheide, 2006).

Although acts of terrorism really do take place—and people really die—terrorism nonetheless can be considered what Boorstin (1961) called a "pseudo-event," something planned for the purpose of being reported by news media. A similar concept is the "media event," introduced by Dayan and Katz (1992) and discussed by Nossek (2008) as a social ritual that builds solidarity by highlighting a nation's meaningful events. Terrorism as media event can trigger a culture's master news narrative of solidarity, partnership, and triumph over the terrorist cause.

Terrorism News as Public Mourning

News coverage of death and disaster represents a form of public mourning that brings society together to share common beliefs, presenting an affirmation of everyday life (Kitch, 2000). It represents a secular, quasi-religious ritual that leads from crisis back to a normal state. Just as news media can eulogize famous public figures, they can eulogize the death of social stability, public safety, and a society divided by calamity. By doing so, media reaffirm group values and identities, bringing the message that leaders are back in charge, that the public is strong, and that ordinary life is becoming, once again, ordinary (Zelizer, 2002). This process can be considered the "pastoral role" of media, highlighting ongoing shared values and renegotiating a shift from a sphere of controversy to a sphere of consensus (Schudson, 2002). In all, this ritual narrative reconnects with a society's ongoing collective memory of itself.

In some cultures, news becomes an opportunity to re-present values through humor and satire. For example, the Danish newspaper *Jyllands Posten* attempted to stretch the boundaries of satire by hosting a collection of cartoons concerning the Muslim prophet Mohammad (Berkowitz & Eko, 2007; Hussain, 2007). Going beyond political cartooning, Zelizer (2002) argued that photography plays a dual role both as a part of journalism and as a tool for easing the trauma of terrorism. In its second role, photography helps a society demonstrate that safety has returned and mourning has been accomplished. Often, the role of a photograph is to validate a news story's text as well (Fahmy, 2007; Huxford, 2001). It is thus crucial for producer and audience to have a shared sense of meaning that reproduces enduring images of historical events, to the point where they become *the expected* images (Griffin, 2004). The candlelight vigil photograph, for example, becomes an expected image following a terrorist attack, communicating shared grief and unity.

Visual Elements, Cultural Meanings

Regardless of whether a visual image is a cartoon, a photograph, or a graphic representation, that image has cultural ties that constrain its intended meaning. In the case of political cartoons, significant decoding is required—understanding context and cultural meanings is crucial to understanding the intent of a political cartoon. Without this guidance, the full intended meaning may never be fully understood (Alkazemi & Wanta, 2015). As Barthes (1977/1985) explains, messages can be denotational or connotational, providing both intuitive meanings and those that are buried deeper in cultural life. In the case of an advertisement for pasta and sauce, for example, images of a green pepper, a red tomato, and yellow pasta help represent "Italianicity" by borrowing the Italian flag's colors in combination with a linguistic, textual message that states the same directly. However, as a "French" message, it plays on certain touristic stereotypes that would barely correspond with the cultural frame of reference of an actual Italian.

As such, Barthes (1977/1985) argued against a universal code of meanings to be derived from any given visual image—instead it can be said that the poster to which Barthes referred contains complementary messages that are linguistic, denoted (the literal) and connotated (the implied). From this tripartite scheme, Barthes's linguistic becomes an "anchorage" that delimits the potential range of meanings that can be drawn from an image.

Rose (2016) used the term "visual culture" to describe how the visual has become part of everyday social life. However, the images of this visual culture are removed from an exact correspondence to the "real world" so that an image-shaped reality becomes only a simulation (a simulacra) of that reality. For the analyst of images in a visual culture, the act of "compositional interpretation" requires "the good eye" that is not explicitly methodological but develops a way of looking that results in a specific sense of understanding images. To apply the good eye effectively requires knowledge of an image's cultural context. Cultural memory comes into play, too, when a viewer applies familiar understanding through the synecdochal, where a part stands in for the whole. Rose used the example of how an image of the Eiffel Tower can serve as a sign for Paris (or even France) as a whole.

In understanding how photographs and cartoons related to terrorist acts convey encoded meanings, it is helpful to reconnect concept to application. For example, a photograph of a memorial gathering that follows a terrorist act would gain anchorage from the photograph's caption, along with linguistic guidance from the news article's headline and text. As an audience member views a photographic image, meanings are negotiated and refined, ideally moving toward the represented meanings that the creator intended. Some of these meanings can be taken more literally and effectively understood, although an audience member with greater cultural understanding can go beyond those basic understandings; a member from outside that visual culture would lack an understanding of dominant codes, would not have the good eye, and would therefore miss many of the meanings contained in those visual signs. In either case, the photograph of one memorial gathering would be synecdochal, representing other similar gatherings, the city overall, the nation, and even Western culture at large.

Decoding Political Cartoons

Compared to a photograph, decoding a political cartoon requires a stronger intuition of the visual culture, the artists' intended connotations and denotations, and in some cases access to the cartoonist's use of language. With these abilities, an audience member would be able to gain an anchorage of the cartoonist's text. This effort requires familiarity with the symbols employed, the ability to read the language used —both literally and figuratively—and knowledge of the nuances of both current and historical events (Hou & Hou, 1998). Cartoons are particularly known as "a performative genre," one that is not expected to represent reality but rather to engage with it and augment it (Hansen, 2011). Although a photographer is expected to represent truth literally, the political cartoonist has the ability exaggerate symbols while applying satire, caricature, and parody to stretch the boundaries of what corresponds to the realm of truth (Diamond, 2002).

One critical challenge of decoding cartoons is that they represent a medium intended for almost instant interpretation. Not only can a cartoon's message be misinterpreted, but without adequate knowledge of relevant historical and cultural cues, a cartoon might not be decoded at all, regardless of the time spent (Hansen, 2011). An obvious illustration of this point is the set of Mohammad cartoons published by Denmark's *Jyllands-Posten*, which could be interpreted as pro-Muslim or anti-Muslim, depending on a reader's vantage point. As visual story elements become parodied in cartoons, this elasticity makes meanings even less concrete (Edwards & Winkler, 1997). The concept of *ideograph*—typically limited to words and phrases such as "liberty" or "patriotism"—also illuminates the meanings that a visual element can represent when it builds off of "an ordinary image" (Edwards & Winkler, 1997, p. 297). For example, the iconic image of Marines raising an American flag on Iwo Jima can have other visual elements juxtaposed in a cartoon—such as a gasoline pump—to carry the ideological meaning of the flag-raising to other ideologically laden issues. To be decoded effectively, however, requires that a viewer be enmeshed into the cartoonist's culture.

To summarize, visual elements such as photographs and cartoons serve as key tools for a society to acknowledge terrorism, share in grief and trauma, and gain solidarity in terrorism's aftermath. These visual elements communicate their messages through both their explicit and implied meanings. Much of what these visuals contain are culturally bound, though, so that their meanings—especially those that are implied—are not always clear to those outside a culture (Darling-Wolf, 2016). Although embedded texts such as photographs of signs or cartoon speech balloons are mainly available to cultural insiders, external texts such as photo captions or the body of a news story can serve as anchorages that help delimit meanings of the visual and make those meanings clearer to cultural outsiders. By engaging in both the textual and the visual, journalism therefore plays an important role, sharing visual elements with a global forum and helping to explain their meanings to their global audiences.

This leads to three research questions:

RQ1: What is the role of photography in the visual representation of solidarity against terrorism?

RQ2: How do political cartoons differ from photography in their visual representation of solidarity against terrorism?

RQ3: How do news texts related to these visual representations help anchor and clarify their cultural meanings?

METHOD

Rose (2016) suggested that semiologists tend not to seek out images that are statistically representative of a wider set of images. Instead, images often are chosen for their conceptual interest and their ability to make a point effectively, which leads to smaller data sets that are scrutinized more closely. She explained that once a set of images is selected, a semiotic analysis can be accomplished through five basic steps (p. 132):

1. Decide what the signs are.
2. Decide what they signify "in themselves."
3. Consider the relation between multiple signs.
4. Explore the connection of signs to broader systems of meaning.
5. Return to the signs to explore their articulation of ideology.

For this study, there was also a goal of learning what a broad range of photographs, cartoons, and their texts indicate about the cultural role of terrorism. For that reason, a set of elements was selected that would help look for patterns both in the types of images and the basic themes about terrorism that they represent. This decision in turn leads to a less fine-grained interpretation of the images, yet one guided by the principles that semiotics provides.

Data for this analysis came from four searches of Google News, beginning with a broad term, "Brussels bombing." These initial results were examined to assess the kinds of images that could be found. From there, more-specific search terms were used that could focus deeper on the initial elements and patterns, including

- "Brussels memorial commemoration," because several photographs of memorial gatherings were found;
- "Brussels fountain cartoon," because many of the cartoons played on memes of the well-known statue Manneken Pis;
- "Tintin terrorist," to explore a multitude of memes based on the famous Belgian cartoon character; and
- "Belgium cartoon," to search out other cartoons about the bombings.

In all, 67 news items from media in the United States, United Kingdom, Australia, Canada, Israel, and the Netherlands were identified from the searches (all appeared in English). Most of these items contained multiple visual elements, with 282 images included. Several of those images had been captured from Twitter or Instagram and

TABLE 1
Summary of Image Themes in News Items Related to the Brussels Bombings

Theme	No. of Images	Description
Aftermath photographs	11	Photographs showing wreckage resulting from the bombings.
Memorial gathering photographs	44	Photographs depicting gatherings of people and the artifacts they brought to the events, such as candles, flags, flowers, and teddy bears.
Building photographs	35	Photographs of buildings illuminated with colors of the Belgian flag or decorated with multiple flags.
Tintin cartoons	81	Cartoon memes and adaptations of the iconic Belgian cartoon character, often shedding tears.
Manneken Pis images	38	Cartoon memes and photographic images of the iconic Belgian statue of a young boy urinating into a fountain.
Flag-related images	52	Cartoons, graphics, and photographs built from the black, yellow, and red vertical bars of the Belgian flag.
Other cartoons	21	Cartoons that do not fit into the other cartoon-related themes, often drawn by artists outside of Belgium.

Note. $N = 282$.

then republished in the news items. Analysis began with a baseline summary of the kinds of elements that were found. After reviewing the images several times, they were grouped into seven thematic categories, as shown in Table 1.

Each of the seven groups was examined for their main themes and the most frequently republished images. Finally, explanatory texts and news articles were analyzed, again looking for the key themes that they addressed. Multiple readings of texts and images together were used to clarify meanings and guide interpretation. Although the news texts and captions were all in English, many of the images included signs, placards, and speech bubbles in other languages, mainly in French. To facilitate a better understanding of these texts, they were run through Google Translate and then reviewed by a student with a degree in French. The English-language news texts that contained translations and explanations were also reviewed to help go beyond the literal meanings of the computer-aided translations. Buildings and monuments such as the Brandenberg Gate in Germany and the Christ the Redeemer statue in Brazil were researched for their meanings, as were foods (french fries, mussels, etc.) and references to other terrorism events (such as bombings in Turkey and Paris).

ANALYSIS

This analysis combines a discussion of the seven categories of images in relation to their anchoring news texts, as a way of placing the texts within their contexts. The image categories appear approximately in the order of least to most interpretive complexity.

Aftermath Photos

These images depicted chaos immediately following the bombings, including damage at an airport luggage area and a heavily damaged train car. Overall, both the images and their explanatory texts were denotative renditions, showing the impact of the bombings. Although no dead bodies were shown, injured or scared people appeared in a few images. Images of rescue workers and authorities, such as police and firefighters, were shown as well.

Because this small set of aftermath photos depicted relatively straightforward subjects, their texts required little cultural explanation to anchor the photos' meanings. Instead, the texts provided basic facts of the attacks. For example, an item from BNO News ("Raw Video," 2016) simply explained, "Three explosions hit the Belgian capital—two at the airport with the third at the city's Maalbeek metro station." A bit of cultural interpretation appeared, however, in an item from *Elite Daily* (Haltiwanger, 2016): "On Tuesday, March 22, terror attacks rocked Brussels, Belgium, one of Europe's most beautiful capitals and the hub of the European Union. There were explosions at the airport and a central train station" (para. 1).

In all, the aftermath images were more or less universal in their representations, requiring little explanation for a media audience to engage in witnessing grief and building solidarity. Their denotative nature required only texts to provide basic facts, such as the number of injuries and fatalities, along with the degree of damage and locations.

Memorial Gathering Photos

As with the photos of the aftermath, photos of memorial gatherings were mainly denotative, requiring little interpretive anchoring by the texts. Common elements of these 44 images were those often found at memorials, as summarized in Table 2.

In all, the meanings of these images were again denotative, so that the texts did not play a big role in cultural translation of their meanings. A typical commentary just added the name of the location where a gathering took place: "Hundreds of people come together at Place de la Bourse to mourn on Wednesday evening, 23 March 2016" (O'Doherty, 2016). Another literal, if more detailed, description appeared on the Dutch news website *NL Times*, along with a translation of a tweet in Dutch:

> Brussels held a small commemoration for the victims of the bombings on the Beirsplein in the city center. Candles were lit and flowers were lain. Hundreds of people visited the site, leaving notes. Belgian Prime Minister Charles Michel visited the site and lit a candle.

TABLE 2
Common Elements of the Memorial Gathering Photographs

Element	Description
Flowers	Appearing singly, in bunches, and in wreaths.
Hearts and peace signs	Appearing on placards, chalked slogans, and other artifacts.
Candles at night	Mainly on the ground at gatherings, sometimes being lit by children and young adults. Candles were sometimes in large groupings, sometimes in the shape of a heart, sometimes in smaller memorials along with a placard.
Crowds and gatherings	People often wearing black clothing, teary-eyed parents with young children, people wrapped in a Belgian flag, people hugging.
People holding placards	Messages about peace, love, and solidarity, often in English.
Chalked messages of solidarity	Often drawn on walls or on the pavement at gatherings. Some written in French, others in English.

So did Jean-Claude Juncker, president of the European Commission. "Solidarity with Belgium. Tonight I'm Belgian." he tweeted. (Peters, 2016)

Short, descriptive captions accompanied a set of 16 photos in Australia's *The New Daily*, for example (Donohoe, 2016):

- "A local Brussels woman holds a paper national flag of Belgium and her daughter a rose."
- "A sign which reads 'we need blood, make a donation' is displayed on the ground near sign as mark of solidarity as people leave tributes at the Place de la Bourse."
- "People light candles at the Place de la Bourse in Brussels."
- "A police officer ignites a candle in honour at Beursplein Square in Brussels."

In the case of these four captions, the general idea is readily accessible, but each adds a few details to anchor nuances of meanings that would not be common knowledge, such as the location of a gathering or a translation of a sign in French concerning blood donations. An image combining several symbols in a Reuters photo (McGrath, Chee, & Fioretti, 2016) fell short of providing the necessary information for a full cultural decoding. Although the caption read, "People gather around a memorial featuring a Menneken-Pis [*sic*] statue in Brussels," only one person's hands and two others' shoes appeared, along with candles, a colorful but unidentified cloth, and a small metal copy of the statue of a urinating boy (with the deeper cultural meaning left aside).

In all, like the aftermath photos, images of memorials required little cultural translation and they easily carried a message of solidarity to the broad media audience. The main challenge in conveying their message involved some non-English signs, which often received translation in the text. Less clear in their meanings were the images involving facsimiles of the Manneken Pis statue, which has served as a long-standing symbol of Belgium and Brussels dating back to 1619.

Photos of Buildings With Colors of the Belgian Flag

On its own, a building illuminated at night with colored lights lacks the anchoring for the images' connotations. A cumulative audience viewing over time similar scenes with different lighting colors, though, could train an audience member to infer that the memorial relates to national colors. However, once the media audience understands the tricolor black–yellow–red flag of Belgium, decoding the images' symbolic solidarity becomes more denotative. Less intuitive, though, was the need to recognize which country stood in solidarity with Belgium—with the notable exception of France, with its iconic Eiffel Tower. The accompanying photo captions and news texts performed another kind of anchoring that clarified the scope of solidarity being conveyed. This included colors of the Belgian flag projected onto these structures, as described in Table 3.

In all, these images of illuminated buildings required two kinds of decoding in order to glean their meanings of solidarity: one to explain the meaning of the

TABLE 3
Location of Buildings and other Structures Illuminated With Colors of the Belgian Flag

Location	Significance
Amsterdam, the Netherlands	The Royal Palace in Dam Square, a magnificent 17th-century building.
Belfast, Northern Ireland	The City Hall, a building that was designed with floodlights that can be set to a variety of colors to commemorate key national holidays and other occasions.
Berlin, Germany	The historic Brandenberg Gate, which has become a symbol of the tumultuous history of Europe and Germany. Also a reminder of European unity.
Brussels, Belgium	The European Union headquarters in the Berlaymont building.
Dubai, United Arab Emirates	The world's tallest building, the Burj Khakifa, known for its magnificent lighting designs.
London, UK	The center of UK government at 10 Downing Street and yet another city that has witnessed terrorist bombings
Paris, France	The Eiffel Tower, one of the best-known landmarks in a country tied to Belgium as both a neighbor and as a recent victim of a similar tragedy.
Rio de Janeiro, Brazil	A huge statue towering over the city, a common symbol of the Rio Olympics and considered a new wonder of the world.
Rome, Italy	Two key city landmarks, the Piazza del Capidoglio and Trevi Fountain.
San Francisco and New York, USA	Major American cities, with photographs of City Hall and One World Trade Center, a location that had experienced a major terrorist attack.

colors, and the other to clarify the monuments' locations. Their meanings were somewhat more connotative than the first two types of images, requiring a higher degree of textual anchoring to aid in their decoding.

Tintin Images

Images and memes adapted from *The Adventures of Tintin* were more symbolically complex than the previous kinds of visuals, requiring decoding at multiple levels. First, a viewer of these images must be aware of the books themselves, as well as the books' central characters and the roles they play in the stories. At a deeper cultural level, a viewer must know that the series of 24 comic books was created by the Belgian cartoonist Georges Remi, who wrote under the pen name Herge (www.tintin.com). A reader of the English translation of these books would likely not be aware of that important detail. Also required for a full cultural decoding of the visual images would be an understanding of the central character, Tintin—a young Belgian reporter and adventurer—and his constant companion, Snowy—a wire fox terrier dog. Tintin and Snowy are often engaged in dangerous situations that require heroic action—an excellent connection between the book series and the nature of the Brussels bombings.

Among the most common images depicting Tintin after the bombings were drawings of Tintin's face with tears in several forms. Some images were just a simple outline drawing of Tintin's head and shoulders with outlined teardrops, often with his hand petting Snowy's head, who also was crying. Some forms of this cartoon image included colored tears that matched the colors of the Belgian flag. One artist drew Tintin's head in black silhouette against a red background, one yellow tear on his cheek. Another variation added a tricolor heart on Tintin's chest. In all, encoding of most Tintin cartoons was highly connotative, requiring a more local understanding of the meanings—globally, these images might not fully resonate with a message of mourning and solidarity.

One particularly prominent Tintin-related image was a color cartoon by Colombian cartoonist Vladdo that appears in Figure 1. In that image, Tintin is wearing his trademark trench coat and reading a newspaper bearing the headline (in French) "Explosions in Brussels." Snowy stands next to Tintin, while he exclaims (again, in French), "My God...!" As with many images, there is a tear on his cheek. A less common yet very compelling cartoon was drawn by South African artist Jerm. In that drawing, Tintin lies on his back, on the pavement in a pool of blood. Snowy stands nearby, as if inspecting him. A black suitcase bears the tag "Brussels."

In all, these cartoon representations were much more connotative in nature, requiring a cultural understanding beyond the literal to truly understand their meanings. The need for some basic language translation—mainly French—takes many of these images yet further from their literal, denotative meanings. As a result, English

FIGURE 1 Belgium's iconic cartoon character, Tintin, with his dog, Snowy.
Note. The black, yellow, and red represent Belgium's flag, and "MORTS" refers to deaths, following Herge's style of using words to highlight key issues and squiggles for the texts. © Vladdo. Reproduced by permission of Vladdo. Permission to reuse must be obtained from the rightsholder.

language texts were essential for building global solidarity. In an article titled "Tintin Cartoon Becomes Social Media Symbol," the HollywoodReporter.com (Richford, 2016) explained,

> Tintin was first created by Belgian artist Herge in the 1920s, and has appeared in various forms in comics and books as well as on radio and television and in film over the decades. The character has become a symbol of Belgium, appearing on postage stamps and coins. (para. 7)

The Australian website and broadcast news outlet 9News ("Brussels Attacks," 2016) added, "Cartoonists around the world have used comic-book hero Tintin and other well-known Belgian icons to mark the series of bomb attacks that have killed around 30 people in Brussels" (para. 1). Meanwhile, *the Toronto Star* (Italie, 2016) directly asserted Tintin's role in creating global unity: "A tearful, beloved cartoon adventurer, Tintin, quickly emerged as a symbol of solidarity in the chaotic aftermath of the Brussels bombings as social media users worldwide took to Facebook and other Web streams" (para. 1). *The Star* further stated,

The books have sold in the tens of millions of copies, but only in Belgium has the fearless reporter and his dog been ingrained in the DNA of most youngsters since the 1950s. Herge died in 1983 and is considered a national treasure in his native Belgium. (para 9)

And *Global News* (Bogart, 2016) concluded, "But, thanks to social media, cartoons are once again becoming a way for people to deal with the horrors of terrorism" (para. 8).

Overall, then, these news texts added cultural meanings to the use of Tintin and his cast of characters that might not be fully apparent to audiences outside of Belgium. Through the cultural interpretation—the anchoring—of the news text in juxtaposition with the cartoons, these images took their essential character from a well-known cultural icon to create a clearer symbol of Belgium's—and the world's —mourning and solidarity, along with a promise of justice.

Manneken Pis Images

As with the images of Tintin, the array of memes involving the statue of a boy urinating—based on the statue Manneken Pis in Brussels—required decoding of both culture and language, with 38 items in all. Without the cultural context, a statue like this would be confusing, possibly even regarded as crude. Put into context by news items, though, the statue became a powerful iconic representation of Belgium's response to terrorism. This was especially the case when cartoons combined with other elements, especially elements that the boy was urinating on, as a signifier of disrespect. These included urinating on a terrorist, on the word *terrorism* (usually in French), on a rifle, and on a bomb. Sometimes, slogans were added, generally in French, including

- "A gift from Belgium. Support for our Belgian friends"
- "Pis and love," which was a play on words, enhanced by a peace sign drawn in the O of the word love.
- "Is a quiet piss too much to ask?"
- "I'll piss on your bombs."

Several objects were added to these basic elements, such as the statue's boy wearing a military helmet and sitting in a sand-bag bunker or having a raised middle finger. In one cartoon, the hand of the Statue of Liberty was holding a cone of french fries in place of the usual torch while the boy urinated on a terrorist with a Belgian flag clenched in his teeth. Another common theme was of the boy crying, sometimes with tears spouting from his eyes like his stream of urine.

Clearly, for media audiences outside of Belgium and nearby countries, media texts were needed to aid decoding of these somewhat obtuse symbolic representations. This was done effectively—if somewhat tongue-in-cheek—by several articles. One piece from the *Huffington Post* (Dicker, 2016) was titled "Brussels Statue is Pissing on Terror All Over Social Media." It went on to state:

Solidarity can take many forms in the wake of a tragedy.

Social media lit up Tuesday with cartoons of a celebrated Belgian statue peeing on terror in the wake of the deadly Brussels attacks, Le Huffington Post reported.

. . .

But the Manneken-Pis, a bronze representation of a little boy urinating into a Brussels fountain, has morphed into a symbol of resilience. (paras. 1–4)

9News ("Brussels Attacks," 2016) added more details to assist in decoding these connotative symbols, explaining that Manneken Pis is one of Brussels's most iconic statues:

'The pissing boy', or 'Mannekin Pis' in Dutch, is a famous bronze fountain in Brussels. The 17th century statue, depicting a boy urinating into a bowl, is only around 60 cm high but draws tens of thousands of visitors every year. (para. 5)

Australia's SBS News ("Social Media Users," 2016) also provided both cultural and historical context for the statue:

Accompanying Tintin in the flood of social media sentiment were altered images of Manneken Pis, a bronze statue of a naked boy urinating into a fountain's basin not far from Brussels Town Hall.

. . .

The boy, dating to around 1619, was created by Brussels sculptor Hieronimus Duquesnoy the Elder. He's the father of sculptor Francois Duquesnoy. The statue has been repeatedly stolen and replaced. It is dressed in costumes on a regular weekly schedule, his wardrobe of several hundred outfits permanently housed in a museum. (paras. 12, 14)

A Reuters article (McGrath et al., 2016) referred to Manneken Pis as a "cheekily urinating boy that has long drawn tourists to Brussels' city center." As the article further explained about details of these memes: "The cherubic Manneken Pis, belovedly irreverent icon of the Bruxellois, was shown relieving himself on a Kalashnikov in images on Twitter using the colors of the Belgian flag: a red boy and yellow gun against a black background" (para. 2).

Without this anchoring contextual information, it would be difficult for an audience member outside of that part of Europe to decode intended meanings of these cartoon memes.

Flag-Related Images

Many of the flag-related images were not completely exclusive from the images of lit-up buildings and the cartoon memes of Tintin and Manneken Pis. In all, 52 items were considered as part of this group. Most notable—and commonly republished— among these images was a cartoon created by Plantu, a longtime contributor to the French newspaper *Le Monde*. That cartoon appears in Figure 2. It shows a line-art character dressed in French flag colors consoling a second, crying character wearing the colors of the Belgian flag. Below each character appears the dates of terror attacks in the two countries. To decode the meanings built into this cartoon, a viewer would need to know the two countries' colors, as well as the relationship between the two countries and their experiences with terrorism.

An examination of flag-related images suggested three main types. The first type was along the lines of the Plantu cartoon just described. One variation of that image appeared in a few versions that incorporated a Turkish character, acknowledging that Turkey also had been a victim of terrorist attacks. A second type of image involved outlines or silhouettes of cartoon characters overlaid or in the background of the Belgian flag. One of these images incorporated a Tintin silhouette, and another

FIGURE 2 Cartoon of solidarity by *Le Monde* artist Plantu.
Note. The character on the left represents France, with its blue, white, and red colors of the French flag. The character on the right represents Belgium, with its black, yellow, and red colors. The dates below the characters stand for the dates of the terrorist attacks in Paris and Brussels. © Plantu. Published in *Le Monde* on March 23, 2016. All rights reserved. Reproduced by permission of Plantu. Permission to reuse must be obtained from the rightsholder.

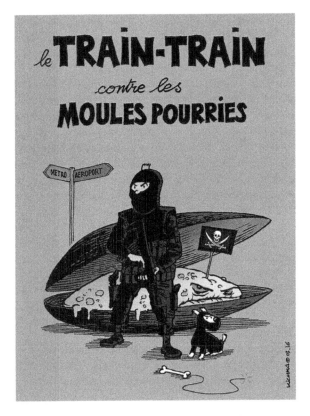

FIGURE 3 Belgium's iconic cartoon character, Tintin, with his dog, Snowy.
Note. In an e-mail, the artist Kichka explained that the cartoon was intended to follow the style of a Tintin book cover. The title of this book translates to *Day-to-Day Routine Life Against the Rotten Mussels*. He portrays Tintin and Snowy as an antiterror cops looking for terrorists, as signified by the rotten mussel behind them. The gray and black colors were intended to create "a sad atmosphere" in contrast to a typical colorful Tintin cover. © Kichka. Reproduced by permission of Kichka. Permission to reuse must be obtained from the rightsholder.

involved the French-Belgian cartoon character Asterix. One more showed a cookie-cutter image of a Smurf created by a splatter painting of the black–yellow–red colors of the Belgian flag (the Smurf is yet another Belgian creation). A third type of image could be considered an element of endearment, which included a blend of the Belgian flag colors with a heart, a peace sign, a crying face, two hands intertwined, and the letters BXL, a common abbreviation for the French name for Brussels, spelled *Bruxelles*.

To successfully decode these three kinds of images would yet again require contextual and cultural knowledge, such as knowing the flag colors; the origins of the cartoon characters; and the history of the French, Belgian, and Turkish terrorist attacks. The *New York Daily News* (Blakinger, 2016), for example, explained the Plantu cartoon in detail, stating that the cartoon offered global solidarity with Belgium: "The illustration shows a crying French flag comforting a tearful Belgian flag. The caption mentions November 13 and March 22—the dates of the attacks in Paris and Brussels, respectively" (para. 2).

Calling the Plantu cartoon "incredibly moving," *Cosmopolitan* (Friedman, 2016) explained that "the French newspaper *Le Monde* posted a tribute from renowned cartoonist Plantu, showing France supporting Belgium in its time of need" (para. 2). The magazine added that the image had more than 5,000 Instagram "likes" and 14,000 Twitter retweets, emphasizing the broad scope of its viewing. *Mic.com* (Noman, 2016) made a similar assertion about the Plantu cartoon:

> Drawings and cartoons circulating on social media also offered catharsis and solidarity in the wake of France's terrorist attacks. Brussels and Paris have been bound by the events in November, as many of the attackers had links to Brussels. Tuesday's suicide bombings will no doubt augment an already strong and suppor-tive relationship between the neighboring countries. (para. 4)

Despite their connection to the Brussels bombings, the other two types of flag-related images—other cartoon characters juxtaposed with the tricolor and endearing images—drew virtually no commentary from texts. This left the images with ambiguity because they were difficult to decode without the necessary textual anchoring.

Other Cartoons

The final type of images that appeared in the news items was a broad collection of cartoons that were drawn by a variety of international cartoonists. This grouping of 21 items required two levels of decoding, one to explain the artists' home country culture and a second to explain the connotative meanings of the specific cartoon. Within this grouping, some items appeared multiple times. One image that appeared four times featured a smoldering man holding up a sign "Je suis Bruxelles" (I am Brussels) while standing in front of a small group of similar-looking characters each holding up an "I am" sign bearing the name of other cities that had also been under terrorist attack. A speech bubble over the man representing Brussels read—in French —"Can you make some room for me?" This cartoon, then, when provided with a contextual interpretation, was another call for solidarity against terrorism. Without the necessary cultural understanding from textual anchoring, these items would be difficult to decode for those outside the cultures.

A cartoon by Belgian Israeli artist Kichka appeared twice in the data set and contained several signs that would be accessible mainly to audience members very familiar with the Belgian culture (Figure 3). Key elements included what at first appeared to be a terrorist holding a rifle and a dog both dressed in black body suits and balaclavas standing in front of a giant, open mussel shell. The French caption translated to "The day-to-day routine against the rotten mussels." On closer exam- ination, the masked characters were actually Tintin and his dog, Snowy. The mussel is a national dish of Belgium, and the creature inside the shell symbolized home- grown terrorism, signified further by a pirate flag. With this decoding, the cartoon could be seen as Belgium standing united against the terrorists lurking within— ultimately suggesting Belgian national solidarity.[1] One other notable image was of a box of french fries, with the middle fry extended outward, with an anchoring English-language caption reading, "Belgium's signature fries appearing to give a one finger salute." In this case, knowing that fries are another national food of Belgium would be essential for decoding—the extended fry "finger" would be much more universal.

Turning to texts related to these images, no texts were provided to help cultural interpretation, other than for the image of the packet of french fries. One article ("Tintin Weeps," 2016) cast the image as a symbol of defiance: "One showed a packet of fries, the country's most famous street food, with one of them seemingly giving the terrorists the finger" (para. 8). A second article, appearing in the United Kingdom's *The Independent* (Payton, 2016) provided a little more detail, "Even Belgium's favourite fried snack, the humble chip, has been used to show defiance towards terror groups such as ISIS and their affiliates" (para. 4). Other than these few comments, the other cartoons were presented more as a gallery, leaving the audience to hold the cultural knowledge to decode the images. Overall, this set of cartoons was the most connotative, requiring adequate textual anchoring for those outside the culture to decode.

DISCUSSION AND CONCLUSIONS

This study began with the notion that visual images—when combined with inter- pretive texts that help anchor their meanings—bring the potential for building global solidarity against terrorism through their messages. The study also argued that some kinds of images—such as photographs of a bombing aftermath or public memorial gatherings—are much simpler to decode and do not rely on texts other than to provide basic details. These more literal images help acknowledge an act of

---

[1] An e-mail exchange with the cartoonist asked who he imagines as his audience. His reply was, "Mainly francophone audience who could understand the codes and symbols. But not only them. I always try to make my work international, as much as I can."

terrorism for a global audience; the audience's general familiarity with terrorism news requires little explicit textual anchoring to make the message understood.

In contrast, editorial cartoons were much more culturally and contextually bound and relied more heavily on anchoring texts to explain language, culture, and facts best known by people relatively close to the situation. Other photographic images— especially illuminated public buildings—required something more, such as the country location of the building and a reminder of the Belgian flag's colors. For example, a photograph of Germany's Brandenberg Gate illuminated with the colors of the Belgian flag—in the context of a news item about the bombings—would symbolize a familiar memorial event in general. Through anchoring texts, though, these kinds of these images gained greater impact and aided the emergence of global messages of solidarity surrounding a specific terrorist act.

The first research question asked about the role of photography in the visual representation of solidarity against terrorism. The photographs that were analyzed placed a particular emphasis on the human experience in a way that invoked empathy and built a connection between media audiences. These images required minimal elaboration because the public memorial gathering has become a common element in media coverage of terrorism in recent years. In some cases, details required explanation when they were culturally specific, such as a small statue of the Manneken Pis or a placard written in a non-English language. Photographs of illuminated buildings—when accompanied by location information—could unite media audiences by showing how commemoration and solidarity went beyond Belgium alone. In the images that were studied, 10 countries were included. Although media audience members would be unlikely to view more than a few different images, seeing even a small number would serve as an effective reminder of the global connection to this terrorist event.

Overall, photographs of terrorism's aftermath, memorials, and color-illuminated buildings gain their effectiveness by enhancing the humanity of terrorism's impact. For a media audience engaged in global issues, then, photographs have the potential to build solidarity—if only vicariously—with a minimal amount of text-enhanced information.

The second research question asked how political cartoons differ from photography in the visual representation of solidarity against terrorism. This study found that, in contrast to photographs, cartoons emphasized how to react to the terrorist attacks, as well as making a concerted statement about terrorists and the act of terrorism itself. Unlike photographs, though, these cartoons required more cultural information to decode. But once explained, a cartoon such as one of Tintin crying became an effective means of creating empathy and building solidarity. In essence, cartoon characters, through their iconic nature, accomplish a degree of synecdoche that becomes more effective than a photograph for expressing a nation's reaction to terrorism. Thus, Tintin represented vigilance and assertive action toward the attacks. The Manneken Pis memes also accomplished this iconic synecdoche while making a blunter statement about solidarity against terrorism than memes of Tintin could without breaking the boundaries of his character. And Plantu's cartoon image of a French flag character comforting a Belgian flag character provided a double dose of synecdoche, representing the solidarity of one nation with another—both of which had faced terrorist attacks in their hearts. In any case, these cartoon images made strong statements but were much more connotative, requiring more cultural/contextual anchoring for an effective reading than would be needed to understand the essence of a photograph. Cartoons, then, would be most effective in conveying meaning to an audience more culturally proximate to a terrorist attack.

The third research question asked how news texts related to these visual representations helped anchor and clarify their cultural meanings. The simple answer is that for all but those living in the European region near Belgium and understanding the French language, English-language news texts would be crucial for providing anchoring to the encoded images. Headlines played an important role of leading media audience members further into a news item to contextualize the visual images and make them more comprehensible. Meanings provided by the combination of text and image gained a chance to promote both empathy and solidarity.

Overall, photographs from this event could be considered more culturally universal, whereas the cartoons contained more localized signs and symbols, offering a more local language and symbols that spoke to those most directly impacted by the bombings. The work of encoding by photographers would be less challenging, whereas the work of encoding—and decoding—messages by cartoonists would be more difficult because of the need to create a more specific, highly symbolic message for their audiences about what to think and what to do. Point of view is the essence of political cartooning. Although political cartoonists have the freedom to embed a variety of messages in their work, photographers are more narrowly bound—a photographer cannot put a flag around the Manneken Pis to construct a photograph, for example, but this kind of symbolism is part of the cartoonist's stock in trade.

Concepts from semiotics here helped highlight the challenge of encoding by producers as well as the differential challenges of decoding photographs versus cartoons. Because of the ongoing terror attacks around the world, media audiences have likely built a visual vocabulary, so that certain kinds of visual images are now understood globally as various culturally unifying responses to terrorism. This suggestion builds on other studies of terrorism imagery, so that the Brussels case joins what other scholars have written. Photographs, in making ritual gestures, communicate how to feel about a terrorist attack. Cartoons, which tend to make instructive and political gestures, go further yet by communicating how media audiences might think and react. Similarly, the visual familiarity of photographs about terrorism speaks outwardly to the rest of the world, whereas cartoons direct their messages to audiences that have more at stake in the wake of a terrorist attack.

Ultimately, the media audience needs to have "the good eye" to meaningfully comprehend intended meanings. When audience decoding does take place successfully, besides providing a stage for the theater of terror, images following terrorist attacks have the potential to help the news media fulfill a pastoral role, creating a forum for healing from grief and communicating solidarity against what terrorism has undone.

ACKNOWLEDGMENTS

The author thanks Tammy Walkner, graduate student at the University of Iowa, for her review of the French translations, and Oren Meyers, University of Haifa, for facilitating contact with cartoonist Kichka.

REFERENCES

Alkazemi, M., & Wanta, W. (2015). Kuwaiti political cartoons during the Arab Spring: Agenda setting and self-censorship. *Journalism: Theory, Practice & Criticism, 16*(5), 630–653.

Altheide, D. (2006). *Terrorism and the politics of fear.* New York, NY: AltaMira.

Barthes, R. (1977 [1985]). Rhetoric of an image. In R. Innis (Ed.), *Semiotics: An introductory anthology* (pp. 192–205). Bloomington: Indiana University.

Berkowitz, D., & Eko, L. (2007). Blasphemy as sacred rite/right: "The Mohammed cartoons affair" and maintenance of journalistic ideology. *Journalism Studies, 8*(5), 779–797.

Blakinger, K. (2016, March 22). Internet shows support for Brussels with emotional Plantu cartoon. *The New York Daily News.* Retrieved from http://www.nydailynews.com/

Bogart, N. (2016, March 24). Iconic Belgian cartoon Tintin sheds tears over Brussels attack. Retrieved from http://globalnews.ca/news/2593237/iconic-belgian-cartoon-tintin-sheds-tears-over-brussels-attack/

Boorstin, D. (1961). *The image: A guide to pseudo-events in America.* New York, NY: Harper & Row.

Brussels attacks: Tintin weeps as cartoonists take on terrorists. (2016, March 23). Retrieved from http://www.9news.com.au/world/2016/03/23/09/02/brussels-attacks-tintin-weeps-as-cartoonists-take-on-terrorists

Darling-Wolf, F. (2016). The lessons of *Charlie*, or locality in the age of globalization. *International Journal of Journalism & Mass Communication, 3,* 115. doi:10.15344/2349-2635/2016/115

Dayan, D., & Katz, E. (1992). *Media events: The live broadcasting of history.* Cambridge, MA: Harvard.

Denham, J. (2015). Paris attacks: Powerful cartoons from around the world. Retrieved from http://www.independent.co.uk/

Diamond, M. (2002). No laughing matter: Post-September 11 political cartoons in Arab/Muslim newspapers. *Political Communication, 19*(2), 251–272.

Dicker, R. (2016, March 22). Brussels statue is pissing on terror all over social media. Retrieved from http://www.huffingtonpost.com/

Donohoe, R. (2016, March 23). Gallery: Beautiful tributes flow for Brussels. Retrieved from http://thenewdaily.com.au/

Edwards, J., & Winkler, C. (1997). Representative form and the visual ideograph: The Iwo Jima image in editorial cartoons. *Quarterly Journal of Speech, 83*(3), 289–310.

El Refaie, E. (2009). Multiliteracies: How readers interpret political cartoons. *Visual Communication*, *8*(2), 181–205.

Fahmy, S. (2007). "They Took It Down": Exploring determinants of visual reporting in the toppling of the saddam statue in national and international newspapers. *Mass Communication and Society*, *10*(2), 143–170.

Fahmy, S. (2010). Contrasting visual frames of our times: A framing analysis of English- and Arabic-language press coverage of war and terrorism. *The International Communication Gazette*, *72*(8), 695–717.

Friedman, M. (2016, March 22). This cartoonist's tribute from Paris to Brussels is incredibly moving. *Cosmopolitan*. Retrieved from http://www.cosmopolitan.com/politics/news/a55574/brussels-attacks-paris-terrorism-plantu-cartoon-tribute/

Griffin, M. (2004). Picturing America's "War on terrorism" in Afghanistan and Iraq. *Journalism: Theory, Practice & Criticism*, *5*(4), 381–402.

Haltiwanger, J. (2016, March 22). France uses this beautiful cartoon to show solidarity with Belgium. Retrieved from http://elitedaily.com/news/french-cartoonist-tribute-brussels-attacks/1429665/.

Hansen, L. (2011). Theorizing the image for security studies: Visual securitization and the Muhammad cartoon crisis. *European Journal of International Relations*, *17*(1), 51–74.

Hou, C. (n.d.). Decoding political cartoons. Retrieved from https://www.collectionscanada.gc.ca/education/008-3050-e.html.

Hou, C., & Hou, C. (1998). *The art of decoding political cartoons: A teacher's guide*. Vancouver, Canada: Moody's Lookout.

Hussain, A. (2007). The media's role in a clash of misconceptions: The case of the Danish Muhammad cartoons. *Press/Politics*, *12*(4), 112–130.

Huxford, J. (2001). Beyond the referential: Uses of visual symbolism in the press. *Journalism: Theory, Practice & Criticism*, *2*(1), 45–71.

Italie, L. (2016, March 22). Crying Tintin cartoon floods social media in aftermath of Brussels attacks. Retrieved from http://www.thestar.com/news/world/2016/03/22/crying-tintin-cartoon-floods-social-media-in-aftermath-of-brussels-attacks.html

Kircher, M. (2016, March 22). French fries are becoming an unlikely symbol of the attacks in Brussels. Retrieved from http://www.businessinsider.com/brussels-terrrorist-attacks-french-fries-instagram-2016-3

Kitch, C. (2000). "A news of feeling as well as fact": Mourning and memorial in American news-magazines. *Journalism: Theory, Practice & Criticism*, *1*(2), 171–195.

Kitch, C. (2002). "A death in the American family": Myth, memory, and national values in the media mourning of John F. Kennedy Jr. *Journalism & Mass Communication Quarterly*, *79*(2), 294–309.

McGrath, M., Chee, F., & Fioretti, J. (2016, March 22). Twitter users express support for Brussels with its iconic fountain. Retrieved from http://www.reuters.com/article/us-belgium-blast-socialmedia-idUSKCN0WO1Z7

Nacos, B. (2016). *Mass-mediated terrorism: Mainstream and digital media in terrorism and counter-terrorism* (3rd ed.). Lanham, MD: Roman & Littlefield.

Noman, N. (2016, March 22). This powerful cartoon has become a symbol of solidarity after the Brussels attacks. Retrieved from http://mic.com/articles/138547/this-powerful-cartoon-has-become-a-symbol-of-solidarity-after-the-brussels-attacks#.dE4unhocc

Nossek, H. (2008). "News media"-media events: Terrorist acts as media events. *Communications*, *33*, 313–330.

O'Doherty, M. (2016, March 23). Belgium terror: Northern Ireland's past proves that dismissing Brussels terrors as "sick" will get us nowhere. *Belfast Telegraph*. Retrieved from http://www.belfasttelegraph.co.uk/

Payton, M. (2016, March 23). Brussels attacks: How cartoonists around the world reacted to the atrocity. Retrieved from http://www.independent.co.uk/news/world/europe/brussels-attacks-h... nists-around-the-world-have-reacted-to-the-tragedy-a6947586.html

Peters, J. (2016, March 23). Netherlands lights up buildings, lowers flags to honor Brussels. Retrieved from http://www.nltimes.nl/2016/03/23/netherlands-lights-up-buildings-lowers-flags-to-honor-brussels/

Raw video: Immediate aftermath inside explosion at Brussels' Zavantem airport. (2016, March 22). Retrieved from http://bnonews.com/news/index.php/news/id3917

Richford, R. (2016, March 22). Brussels attacks: Tintin cartoon becomes social media symbol. *The Hollywood Reporter*. Retrieved from http://www.hollywoodreporter.com/news/brussels-attacks-tin tin-cartoon-becomes-877308

Rose, G. (2016). *Visual methodologies: An introduction to researching with visual materials* (4th ed.). Thousand Oaks, CA: Sage.

Schudson, M. (2002). What's unusual about covering politics as usual. In B. Zelizer & S. Allan (Eds.), *Journalism after September 11* (pp. 36–47). New York, NY: Routledge.

Social media users share crying Tintin. (2016, March 23). Retrieved from http://www.sbs.com.au/news/article/2016/03/23/social-media-users-share-crying-tintin

Tintin weeps for Belgium but forgets Turkey. (2016, March 22). Retrieved from http://www.worldbulletin.net/headlines/170777/tintin-weeps-for-belgium-but-forgets-turkey

Weimann, G. (1983). The theater of terror: Effects of press coverage. *Journal of Communication, 33* (1), 38–45.

Zelizer, B. (2002). Photography, journalism, and trauma. In B. Zelizer & S. Allan (Eds.), *Journalism after September 11* (pp. 48–68). New York, NY: Routledge.

Graphic Violence as Moral Motivator: The Effects of Graphically Violent Content in News

Matthew Grizzard (Ph.D., Michigan State University, 2013) is an Assistant Professor in the Department of Communication at the University of Buffalo, The State University of New York. His research interests include moral emotions and moral judgment processes related to the consumption of narrative and interactive media entertainment.

Jialing Huang (M.A., University of Miami, 2014) is a Ph.D. student in the Department of Communication at the University at Buffalo, The State University of New York. Her research interests include media entertainment and media effects.

Julia K. Weiss (M.A., University at Buffalo, The State University of New York, 2015) is a Ph.D. student in the Department of Communication Studies at West Virginia University. Her research interests include emotion-based perspectives of media effects and persuasion.

Eric Robert Novotny (M.A., University at Buffalo, The State University of New York, 2015) is a Ph.D. student in the Department of Communication at Michigan State University. His research interests include interpersonal motor synchrony and its ability to elicit positive social outcomes as well as virtual reality and motion capture technology as a means of measuring synchrony and its outcomes in a controlled environment.

Kaitlin S. Fitzgerald (M.A., University at Buffalo, The State University of New York, 2017) is a Ph.D. student in the Department of Communication at University at Buffalo, The State University of New York. Her research interests include narrative influence and engagement, particularly within the context of entertainment media.

Changhyun Ahn (M.A., Cleveland State University, 2015) is a Ph.D. student in the Department of Communication at University at Buffalo, The State University of New York. Her research interests include emotional and cognitive processing of video games.

Zed Ngoh (M.A., University at Buffalo, The State University of New York, 2016) is a graduate from the Department of Communication, University at Buffalo and a data analyst in a private organization. His research interests include the psychological effects of entertainment media.

Alexandra Plante (M.A., University at Buffalo, The State University of New York, 2016) is the Director of Communications for the Recovery Research Institute at Massachusetts General Hospital and Harvard Medical School. Her research interests include translating addiction science through media and narrative.

Haoran Chu (M.A., Cardiff University, 2014) is is a Ph.D. student in the Department of Communication at University at Buffalo, The State University of New York. His research interests include climate change communication.

Matthew Grizzard and Jialing Huang

Julia K. Weiss

Eric Robert Novotny

Kaitlin S. Fitzgerald and Changhyun Ahn

Zed Ngoh

Alexandra Plante

Haoran Chu

Common wisdom holds that graphic media violence leads to antisocial outcomes. This common wisdom is reflected in the Society for Professional Journalists' Code of Ethics. However, theory and research regarding moral emotions' ability to increase moral sensitivity suggests that this type of negative content may be capable of yielding prosocial responses. This article describes this logic and tests its predictions in two experimental studies utilizing news footage of a mass execution conducted by the Islamic State in Iraq and Syria (ISIS). Results corroborate claims that graphic media violence can serve as a moral motivator. Higher levels of graphic violence led to stronger anger and disgust responses, which in turn predicted higher levels of (a) moral sensitivity, (b) desires for anti-ISIS interventions (including military and humanitarian efforts), and (c) eudaimonic motivations (i.e., seeking meaning in life). Important to note, no increases in negative attitudes toward Arab Muslims were observed. Theoretical implications are discussed.

Technological convergence has enabled terrorist organizations to produce and distribute high quality propaganda videos featuring extreme acts of violence (Carr, 2014; Grizzard, 2016). News media often cover the violent activities present within these videos (e.g., mass executions) but struggle with the ethical ramifications of displaying highly graphic images of the acts (see Fahmy, Bock, & Wanta, 2014). Displaying graphic violence in news can result in claims of sensationalism (see Hoffner et al., 1999; van der Molen, 2004; Vettehen, Nuijten, & Beentjes, 2005) and the alienation of viewers (Keith, Schwalbe, & Silcock, 2006). At the same time, failing to display graphic violence can stymie accurate reporting (Fahmy et al., 2014) and occlude the true consequences of human tragedies (Cook, 2001; Fahmy, 2005; Kratzer & Kratzer, 2003; McKinley & Fahmy, 2011). Although this is a debated topic within journalism, official statements from the Society of Professional Journalists (SPJ) guide journalists away from displaying graphic violence by implying that its display is unnecessary and sensationalistic. For example, SPJ's position paper "Reporting on Grief, Tragedy, and Victims" (Purmalo, n.d.) addresses coverage of violence and crime, and it implores journalists to "recognize that gathering and reporting information may cause harm or discomfort" (para. 8) and "show good taste. Avoid pandering to lurid curiosity" (para. 10). It, moreover, reassures journalists that "news [of grief and tragedy] will draw attention no matter the presentation. In other words, media will receive higher marks if they present the stories in responsible fashion without resorting to sensationalism in words or photos" (para. 14). Although these statements reflect common wisdom that media violence has predominantly antisocial effects, recent research has challenged the validity of past research supporting this position and called for an impartial examination of potentially prosocial effects of violent media (see Elson & Ferguson, 2014; Ferguson, 2010, 2015; Markey, Males, French, & Markey, 2015). In fact, some theories predict that displaying graphic violence may enhance attention and amplify desires to quell violence and aid victims (Fahmy et al., 2014; Grizzard, 2016; Scharrer & Blackburn, 2015).

MORAL PSYCHOLOGY AND MORAL FOUNDATIONS THEORY

Current theories within moral psychology conceptualize moral judgments (i.e., categorizing an act as good or bad) as primarily resulting from intuitive and

31

emotional processes (Haidt & Joseph, 2008). From these perspectives, moral judgments result from affective responses to stimuli, which lead to heuristic judgments (Haidt & Joseph, 2008). Such affective responses are referred to as *moral emotions* (Haidt, 2003; Tangney, Stuewig, & Mashek, 2007). Moral emotions are typically considered in terms of a self-other dimension and an upholding-violating dimension (Tangney et al., 2007): Moral violation by the self is likely to elicit guilt or shame, whereas upholding is likely to elicit moral pride. Witnessing another violate morality is likely to elicit contempt, anger, and disgust, whereas witnessing another uphold morality is likely to elicit elevation. Researchers have begun to explicate the types of actions that will stimulate moral emotions in broad integrated theories. One such theory linked to media effects (see Tamborini, 2013) is moral foundations theory (MFT; Haidt & Joseph, 2008).

According to MFT (Haidt & Joseph, 2008), evolution has resulted in humans' possessing sensitivities and biases toward specific behaviors. These sensitivities are described as the five moral intuitions: care (related to inflicting emotional or physical harm on others), fairness (related to justice concerns and the breaking of social contracts), loyalty (related to ingroup biases and allegiances to one's own group), authority (related to hierarchies and deference to legitimate authority), and purity (related to bodily sanctity and the avoidance of engaging in unclean or disgusting actions; see Graham et al., 2011; Haidt & Joseph, 2008). Of import, these five moral intuitions allow for categorization of behavior in terms of upholding/violating a specific moral intuition. For example, giving aid to an injured person would be an upholding of care, whereas injecting heroin would be a violation of purity.

Although MFT assumes that the five moral intuitions are innate and evolutionarily derived, the importance each intuition holds can vary between and within individuals, and this variance is referred to as *moral intuition sensitivity*. For example, some individuals do not view recreational drug usage in moral terms at all; it is neither moral nor immoral. Others view it as highly immoral. Variance in purity sensitivity would be expected to predict whether one viewed drug usage as immoral and the magnitude of the immorality judgment. Past research on a large international sample indicates that variance in moral intuition sensitivity is associated with culture, political affiliation, occupation, and religious attendance (Graham et al., 2011), suggesting that moral intuition sensitivity is somewhat stable and akin to a trait. However, because moral judgments are tied to emotions, which are inherently transient states, moral intuition sensitivity is also subject to temporary enhancement.

Media psychology research indicates that media exposure can increase moral intuition sensitivity through priming and social learning mechanisms (Eden et al., 2014; Grizzard, Tamborini, Lewis, Wang, & Prabhu, 2014; Tamborini, 2013; Tamborini, Prabhu, Lewis, Grizzard, & Eden, 2016). Moreover, some of this research indicates that the influence of media exposure on increased sensitivity is mediated by the elicitation of moral emotions (Grizzard et al., 2014). In an experimental study, Grizzard et al. (2014) had participants play a violent video game under two conditions: a guilt-inducing condition (playing as a terrorist) and a non-guilt-inducing condition (playing as a United Nations soldier). Following game play, guilt and moral sensitivity were assessed. Results were consistent with a mediating role of emotion and indicated that playing as a terrorist elicited higher levels of guilt, which in turn elicited greater moral sensitivity toward the moral intuitions violated during game play (i.e., care and fairness). These findings are consistent with emotional models of moral judgment (see Prinz & Nichols, 2010), whereby moral emotions intensify moral judgment processes (Tangney et al., 2007). If emotions intensify moral judgment, then theories that focus on the emotionality of media content (e.g., exemplification theory; Zillmann, 2002) might provide relevant theoretical propositions.

EXEMPLIFICATION THEORY

Exemplification theory (Zillmann, 2002) predicts that exposure to case reports within media content (i.e., exemplars or specific instances) will elicit stronger responses than exposure to abstract information, such as statistics. This broad prediction is based on two primary assumptions relevant to humans' cognitive processing (see Zillmann, 2002, pp. 25–26). First, the theory assumes that concrete observable occurrences place "fewer demands on cognitive processing" (p. 25) than abstract events and, as such, are easier to comprehend, store, and retrieve from memory. Second, consequential events that elicit emotional responses "attract more attention and are more vigorously processed" (p. 26) than nonemotional events. Combined, these assumptions suggest that although visual depictions should elicit stronger effects than text-based depictions (Zillmann, 2002), the more emotional and concrete (i.e., graphic) a visual depiction, the stronger its effects. These predictions have been examined and confirmed in several studies. For example, Zillmann and Gan (1996; as cited in Zillmann, 2002) found that more graphic images of skin cancer in a news story about the risks of sun tanning led to greater assessments of risk than more sanitized images (see Borland et al., 2009, for a similar longitudinal study on graphic-image warning labels on cigarettes). A more related study showed that higher levels of graphicness in images of Israeli–Palestinian conflict resulted in increased negative affect (McKinley & Fahmy, 2011). These findings indicate that more graphic images can lead to stronger emotional responses.

more graphic, stronger reaction (handwritten margin note)

CURRENT STUDIES

Moral emotion research suggests that witnessing a moral violation elicits the moral emotions of contempt, anger, and disgust (Hutcherson & Gross, 2011; Prinz & Nichols, 2010; Tangney et al., 2007). Exemplification research indicates that more graphic content will elicit stronger emotional responses than less graphic content (Zillmann, 2002). In the case of viewing a moral violation, these combined predictions suggest that more graphic depictions of a moral violation will elicit higher levels of contempt, anger, and disgust as compared to less graphic depictions (H1). Furthermore, if the elicitation of moral emotions is a motivating force for moral judgment (see Grizzard et al., 2014; Prinz & Nichols, 2010), then more graphic depictions of moral violations should have an indirect positive effect on moral sensitivity mediated by moral emotions (H2). These hypotheses challenge the conclusions of SPJ's article covering the display of graphic violence. Rather than suggesting that news of violence "will draw attention no matter the presentation" (Pumarlo, n.d., para. 14), exemplification theory and moral psychology suggest that the visuals included in news guide public perception.

Additional Variables and Hypotheses

In addition to moral sensitivity, we also measured several other dependent variables implicated by moral emotion research and exemplification theory. Theory and research on these variables suggest that they too should be amplified by emotional responses in a similar manner to moral sensitivity.

First, exemplification theory focuses on perceptions of risk as a relevant outcome (see Aust & Zillmann, 1996; Zillmann, 2002). Although prior research showed that high levels of graphicness did not lead to perceived severity of the issue or attitudes toward U.S. intervention in international conflicts (McKinley & Fahmy, 2011), the effect may be indirect through the elicitation of emotions, consistent with moral psychology research. Thus, emotional responses (i.e., contempt, anger, and disgust) should positively predict perceptions of risk associated with the elicitor of the emotion (in this case, the perpetrator of violence; H3). Second, research suggests that moral emotions motivate a desire for moral behavior (Hutcherson & Gross, 2011; Prinz & Nichols, 2010; Tangney et al., 2007). In the current case, this desire should relate to preferences for interventions designed to quell the violence and comfort the victims (e.g., military interventions targeting the terrorists and humanitarian interventions aiding their victims; H4). Third, prior research has shown that observing moral upholding can increase *eudaimonic* motivations (i.e., motivations related to being a better person and seeking meaning in life; Krämer et al., 2016; Oliver, Hartmann, & Woolley, 2012). This research linked the elicitation of *positive* moral emotions (e.g., elevation) with these effects. At the same time, eudaimonic motivations have also been linked to graphic violence—a moral violation (Bartsch & Mares, 2014; Bartsch et al., 2016). Combined, these findings suggest that perhaps eudaimonic motivations are elicited by moral emotions regardless of their valence. Thus, we predict that moral emotions elicited by graphic violence will increase eudaimonic motivations (H5). Finally, although it is possible for depictions of terrorist victims to elicit positive moral responses (e.g., sympathy), these depictions might also elicit antipathy toward the perpetrators of such violence. Notably, this antipathy might extend not just to perpetrators but also to the cultural groups from which they originate. Indeed, previous research indicates that media coverage of terrorism can increase prejudice (Das, Bushman, Bezemer, Kerkhof, & Vermeulen, 2009; Lichtblau, 2015). Therefore, our replication included as dependent variables two measures of attitudes toward Arab Muslims—an explicit measure of Islamophobia, which we expect to be positively predicted by moral emotions (H6a), and an implicit measure of attitudes toward Arabs, which we also expect to be positively predicted by moral emotions (H6b). We used both explicit and implicit measures because social desirability may lead participants to underreport such attitudes in explicit measures (see Kaplan, 2006); implicit measures are purportedly less susceptible to such biases (see Greenwald, Nosek, & Banaji, 2003; see also Oswald, Mitchell, Blanton, Jaccard, & Tetlock, 2013, for a critique).

METHOD

Overview and Procedure

We conducted two studies examining whether exposure to more graphically violent news coverage of a mass execution conducted by the Islamic State of Iraq and Syria (ISIS) might increase moral responses to the story. Study 1 (data collection from March 4–12, 2015) was a single-factor between-subjects experiment with four conditions (high, medium, and low graphic violence news story, and an offset control news story that contained no violence). Study 2 (data collection from October 19–27, 2015) was a direct replication of Study 1 to determine the replicability of the effects found in Study 1 and whether graphic violence might increase antipathy toward Arab Muslims.

The procedure was similar for both studies. Participants arrived at the lab and informed consent was obtained. Next, each participant was seated in front of a computer and watched an environmental news story about migrating birds and answered questions about the news story. This story was included purely as a buffer to familiarize participants with the procedure. Participants were then randomly assigned by the experimental software to view one of four videos edited to vary in the level of graphic violence (Study 1 ns: high = 90, medium = 78, low = 66, control = 81; Study 2 ns: high = 67, medium = 63, low = 67, control = 65).

Although the sample sizes for Study 1 were not equal, their deviation did not exceed that which was expected by chance, $\chi^2(3) = 3.74$, $p = .29$. Study 2 indicated a similar nonsignificant chi-square value for deviation from chance, $\chi^2(3) = 0.17$, $p = .98$. Random assignment for the two studies appeared successful. Participants then completed the dependent measures and were debriefed. After reviewing all stimuli, the procedures of the study were determined to be exempt by the Institutional Review Board where data were collected on February 23, 2015, under Exemption 2.

Sample

Participants ($N_{Study1} = 315$, $N_{Study2} = 262$) were recruited from entry-level communication courses at a public university in the northeastern United States and received class credit as compensation: Participants in Study 1 ($n_{male} = 173$, $n_{female} = 140$, $n_{undisclosed} = 2$; $M_{age} = 19.94$, $SD = 2.87$, range $= 18–50$; White/Caucasian, $n = 147$, 46.7%; Asian, $n = 115$, 36.5%; Black/African American, $n = 20$, 6.3%; no other race > 5% of the sample) and participants in Study 2 ($n_{male} = 138$, $n_{female} = 124$; $M_{age} = 20.09$, $SD = 1.92$, range $= 18–32$; White/Caucasian, $n = 120$, 45.8%; Asian, $n = 84$, 32.1%; Black/African America, $n = 28$, 10.7%; no other race > 5% of the sample).

Experimental Stimuli

Our experimental stimuli were based on a news story sourced from a major American broadcast network. The story covered a mass execution conducted by ISIS on Iraqi victims in two scenes: ISIS militants (a) lining victims up in a ditch and shooting them with automatic rifles, and (b) marching victims toward a river before shooting them in the head with pistols and dumping their bodies in the river. The version of the story that was aired in the United States was utilized as our medium-graphicness stimulus, which partially sanitized the violence by freezing the image prior to the weapons' firing; after the image froze, the guns could be heard firing but were not depicted visually. From this version, we created high- and low-graphicness stimuli. In the high-graphicness clip, we replaced the freeze-frame segment with original footage released by ISIS, so instead of the image freezing, the deaths of some of the victims were displayed. In the low-graphicness clip, we replaced the executions with images of the victims being driven to the execution site in cattle trucks. As in the other clips, the deaths were described, but no images of victims at the site of the execution were included. Note that in all versions, the audio, length, and all other aspects remained the same. The only difference between versions was the extent to which the victims' deaths were depicted. As a no-exposure control, we utilized a news story featuring the same anchor and similar length that described neural damage experienced by professional football players. No violent or graphic images were included.

Measures

Moral Emotions. To measure emotional responses, we used Hutcherson and Gross's (2011) procedure. Participants were presented with the prompt, "Please indicate how much you feel each emotion when you think about the news segment," followed by a list of emotions. The moral emotion items were "contempt," "angry," and "morally disgusted" (Hutcherson & Gross, 2011; Tangney et al., 2007). We also measured 12 other emotions (e.g., happy, sad, excited) to disguise the three primary emotions of interest. Participants responded to each emotion on a 7-point response scale, from 1 (*not at all*) to 7 (*extremely*). Following this, participants responded to the prompt, "Please select the emotion that best describes your overall reaction to the news segment." Presentation order of the emotions was randomized. In both studies, scores on contempt were very low, and the modal response was "not at all" (37% of the participants in Study 1 and 45% in Study 2, suggesting that a plurality of participants did not feel contempt at all). In addition, contempt did not correlate strongly with anger ($r_{Study\ 1} = .11$, $r_{Study\ 2} = .11$) or moral disgust ($r_{Study\ 1} = .14$, $r_{Study\ 2} = .12$). Anger and moral disgust, however, correlated highly ($r_{Study\ 1} = .80$, $r_{Study\ 2} = .81$). This pattern suggests that perhaps our participants did not recognize or understand the emotion of contempt. It is also possible—as the research by Hutcherson and Gross (2011) indicated—that although anger and disgust are elicited by moral violation, contempt is elicited by incompetence. Based on these facts, we created an anger-disgust composite by averaging responses to these two items ($\alpha_{Study\ 1} = .90$, $\alpha_{Study\ 2} = .89$).

Moral Intuition Sensitivity. Consistent with prior research (Eden et al., 2014; Grizzard et al., 2014; Joeckel, Bowman, & Dogruel, 2012), we measured sensitivity to the five moral intuitions with the Moral Foundations Questionnaire (Graham et al., 2011). The Moral Foundations Questionnaire employs three relevance items (7-point scale from 1 [*not at all relevant*] to 7 [*extremely relevant*]) and three statement items (Likert-type 7-point scale) for each of the five moral intuitions. Sample relevance items include, "Whether or not someone . . ." "suffered emotionally" (care: $\alpha_{Study\ 1} = .60$, $\alpha_{Study\ 2} = .70$) and "acted unfairly" (fairness: $\alpha_{Study\ 1} = .65$, $\alpha_{Study\ 2} = .76$). Sample statement items include, "People should be loyal to their family members, even when they have done something wrong" (loyalty: $\alpha_{Study\ 1} = .62$, $\alpha_{Study\ 2} = .68$); "Respect for authority is something all children need to learn" (authority: $\alpha_{Study\ 1} = .59$, $\alpha_{Study\ 2} = .60$); and "People should not do things that are disgusting, even if no one is harmed" (purity: $\alpha_{Study\ 1}$ = .66, $\alpha_{Study\ 2}$ = .72). Internal consistency, although somewhat low, is consistent with prior research (see Eden et al., 2014).

Perceptions of Risk from ISIS.   Six questions based on previous research (e.g., Aust & Zillmann, 1996; Zillmann, 2002) assessed perceptions of the risk posed by ISIS: (a) "In your opinion, how serious a national problem is ISIS?" (b) "How likely is it that ISIS will become a problem where you live?" (c) "How likely is it that you personally might become a victim of ISIS?" (d) "How likely do you think it is that the ISIS problem will get worse in the future?" (e) "How likely do you think it is that ISIS will be destroyed?" and (f) "How likely is it that ISIS will conduct another mass execution in one month?" Participants responded on a 9-point scale, ranging 1 (*not at all*) to 5 (*moderate*) to 9 (*very much*). To reduce data, we conducted a principal component analysis with promax rotation (κ = 4). Results from Study 1 indicated three factors best fit the data (76.15% explained variance). Factor 1 (*ISIS as a Problem*; α = .73) included Items 1, 4, and 6; pattern matrix factor loadings on the factor ranged from .70 to .87. Factor 2 (*ISIS as Personal Threat*; α = .71) included Items 2 and 3; pattern matrix loadings on the factor were .86 and .89, respectively. Factor 3 (*ISIS Destruction*) included Item 5 with a factor loading of .98. No item loaded on another factor greater than ± .20. We conducted the same principal component analysis on Study 2's data, restricting the number of factors to three for consistency with Study 1. Results indicated all items loaded on the same factors (81.9% explained variance). The factor loadings for Factor 1 (α = .82) ranged from .83 to .89. The factor loadings for Factor 2 (α = .80) ranged from .90 to .93. The factor loading for Factor 3 was .99. No item loaded on another factor > ± .16.

Desire for Interventions.   We measured participants' opinions regarding increasing U.S. military actions and humanitarian efforts using single-item measures with a 9-point response scale ranging from 1 (*not at all*) to 5 (*moderate*) to 9 (*very much*): "How much do you think the U.S. government should increase military actions (e.g., conducting airstrikes and increasing U.S. forces, etc.) to combat ISIS?" and "How much do you think the U.S. government should increase the humanitarian effort (e.g., sending meals and drinking water to refugees) in Iraq?"

Motivations.   We utilized a scale from prior research (Oliver et al., 2012; see also Krämer et al., 2016) to measure eudaimonic motivations as well as hedonic motivations (i.e., motivations related to pleasure and materialism). The scale consisted of the prompt, "After watching these news stories, I want to ..." followed by five items assessing eudaimonic motivations (e.g., "Do good things for other people;" $\alpha_{Study\ 1}$ = .92, $\alpha_{Study\ 2}$ = .93) and six hedonic/noneudaimonic motivations (e.g., "Meet new people," "Enjoy myself"; $\alpha_{Study\ 1}$ = .88, $\alpha_{Study\ 2}$ = .89). We added two other motivations related to physical health (i.e., "Be healthier" and "Take better care of myself"; $\alpha_{Study\ 1}$ = .88, $\alpha_{Study\ 2}$ = .83). The response options were along a 7-point Likert-type scale.

Attitudes Toward Arab Muslims. Finally, in Study 2, we measured implicit and explicit attitudes toward Arab Muslims. We measured implicit attitudes toward them using the Arab-Muslim Implicit Association Test (A-M IAT 2015; Park, Felix, & Lee, 2007), retrieved from the Inquisit test library and executed using Inquisit 4 (see version 4; Agerström & Rooth, 2009; Park et al., 2007). Participants sorted names that are either "Arab-Muslim" (e.g., Yousef) or "Other People" (e.g., Philippe) and affectively valenced words that were either "Good" (e.g., Joy) or "Bad" (e.g., Awful). By crossing the name categories with the word categories, the A-M IAT assesses the strength of the association between "Arab-Muslim = Bad" and "Other People = Good." Scores on the A-M IAT can theoretically range from −2 to +2 with positive scores indicating an "Arab-Muslim = Bad, Other people = Good" bias and negative scores indicating an "Arab-Muslim = Good, Other People = Bad" bias. Regarding the "Arab-Muslim Implicit Attitude Test," it deserves noting that the measure specifically assesses attitudes toward Arabs who are Muslims (not all Arabs are Muslims and the majority of Muslims are not Arabs). Scores in our sample ranged from −1.34 to 1.25 with a mean of 0.14 (SD = 0.44), which was significantly more positive than a score of 0, $t(261)$ = 5.04, p < .001. Following the implicit measure, we employed the cognitive subscale of the Islamophobia Scale to measure explicit Islamophobic attitudes (Lee, Gibbons, Thompson, & Timani, 2009). This subscale consists of eight items (e.g., "Islam is a dangerous religion"; α = .97) with a 7-point Likert-type response scale. To assess the measurement validity, we correlated both the explicit and implicit scales with demographic data. The 16 participants who identified as Muslim had lower scores on both the explicit (M = 1.72, SD = 1.66), $t(260)$ = 3.14, p = .002, Cohen's d = 0.39, and implicit measures (M = −0.10, SD = 0.45), $t(260)$ = 2.25, p = .03, Cohen's d = 0.28, as compared to non-Muslim identifying participants ($M_{Explicit}$ = 2.92, SD = 1.47, $M_{Implicit}$ = .15, SD = 0.44). Implicit and explicit measures were nonsignificantly correlated (r = -.07, p = .28).

RESULTS

We submitted data from both Study 1 and Study 2 to the same analysis techniques. Results are reported together to provide a comparison of the findings. We first examined the histograms to determine whether the amount of anger-disgust differed between conditions (H1; see Figure 1). As expected, the control

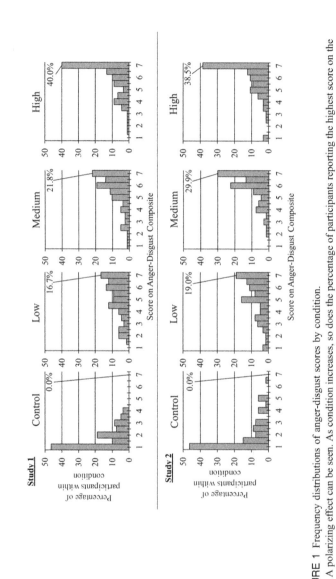

FIGURE 1 Frequency distributions of anger-disgust scores by condition.

Note. A polarizing effect can be seen. As condition increases, so does the percentage of participants reporting the highest score on the anger-disgust composite (i.e., seven of seven).

stimulus elicited very low levels of anger-disgust, whereas the three experimental conditions elicited high levels. Because (a) the inclusion of the control condition could inflate results from statistical tests due to the low levels of anger-disgust in this group and (b) the most informative comparisons are between the low-, medium-, and high-graphicness experimental conditions, we limited our significance testing to these conditions. Results of a linear regression indicated that condition (dummy-coded linearly, -1 = low, 0 = medium, $+1$ = high) was a significant predictor of anger-disgust in both studies ($\beta_{Study\ 1}$ = .21, p = .001; $\beta_{Study\ 2}$ = .19, p = .007). Residuals deviated from normal. To ensure that our results were not simply a statistical artifact, we recoded anger-disgust as a binary variable (i.e., maximum score, 7 = 1, and all other scores = 0) and subjected the recoded data to a logistic regression with dummy-coded condition as the predictor. The model was significant for both Study 1, $\chi^2(1, N = 234) = 11.53$, $p = .001$ ($B_{Study\ 1} = 0.64$, $SE = 0.20$, Wald $\chi^2 = 10.69$, $p = .001$, odds ratio = 1.90), and Study 2, $\chi^2 (1, N = 195) = 5.89$, $p = .02$ ($B_{Study\ 2} = 0.48$, $SE = 0.20$, Wald $\chi^2 = 5.70$, $p = .02$, odds ratio = 1.62). Thus, the robustness check was consistent with the linear regression results.

To examine whether anger-disgust would directly predict the dependent variables and mediate the effect of condition on the dependent variables (H2–H6), we used Hayes's PROCESS macro (version 2.16; Hayes, 2013). We ran a series of mediation models (Model 4) with bias-corrected confidence intervals (10,000 bootstrap samples). Condition predicted anger-disgust (Path a), which in turn predicted the dependent variables (Path b). We assessed mediation by examining the significance of the indirect effect (i.e., did the 95% confidence interval cross zero?). Results of the mediation analyses were largely consistent with our hypotheses (see Table 1). Across all models, graphic violence was a significant positive predictor of anger-disgust.

With regard to moral intuition sensitivity (H2), anger-disgust was positively associated with greater sensitivity of all five moral intuitions (care, fairness, loyalty, authority, and purity) in both studies. The indirect effect was significantly different from zero, suggesting that feelings of anger-disgust mediate the influence of graphic violence on moral sensitivity (see Table 1).

With regard to risk perceptions of ISIS (H3), anger-disgust positively predicted perceptions of ISIS as a Problem in both studies, and the significant indirect effect was consistent with mediation. These findings suggest that higher levels of graphic violence increase anger-disgust, which in turn leads to stronger perceptions of ISIS as a Problem. Anger-disgust did not significantly predict ISIS as Personal Threat in either study. These findings indicate that higher levels of graphic violence do not lead to stronger perceptions of personal risk associated with ISIS. Finally, with regard to ISIS Destruction, the findings of the two studies diverge. Anger-disgust did not positively predict perceptions that ISIS would be destroyed in Study 1 but did in Study 2. In addition, the results of the

TABLE 1
Results From Mediation Analyses

Outcome Variable	Path a	Path b	Indirect Effect (95% Confidence Interval)
Care			
Study 1	.38*** (.12)	.17*** (.04)	.07* (.03, .13)
Study 2	.35** (.14)	.26*** (.05)	.09* (.02, .18)
Fairness			
Study 1	.38*** (.12)	.18*** (.03)	.07* (.03, .13)
Study 2	.35** (.14)	.29*** (.05)	.10* (.03, .21)
Loyalty			
Study 1	.38*** (.12)	.11** (.04)	.04* (.01, .10)
Study 2	.35** (.14)	.15** (.05)	.05* (.01, .13)
Authority			
Study 1	.38*** (.12)	.16*** (.04)	.06* (.02, .13)
Study 2	.35** (.14)	.16*** (.05)	.06* (.01, .14)
Purity			
Study 1	.38*** (.12)	.20*** (.05)	.08* (.03, .15)
Study 2	.35** (.14)	.17*** (.06)	.06* (.01, .15)
ISIS as Problem			
Study 1	.38*** (.12)	.35*** (.07)	.14* (.05, .26)
Study 2	.35** (.14)	.50*** (.08)	.18* (.05, .34)
ISIS as Personal Threat			
Study 1	.38*** (.12)	$.03^{ns}$ (.07)	$.01^{ns}$ (-.04, .07)
Study 2	.35** (.14)	$.17^{ns}$ (.10)	.06* (.00, .18)
ISIS Destruction			
Study 1	.38*** (.12)	$.05^{ns}$ (.09)	$.02^{ns}$ (-.04, .11)
Study 2	.35** (.14)	.31** (.11)	.11* (.03, .26)
U.S. Military Intervention			
Study 1	.38*** (.12)	.37*** (.09)	.14* (.05, .28)
Study 2	.35** (.14)	.45*** (.11)	.16* (.05, .34)
U.S. Humanitarian Intervention			
Study 1	.38*** (.12)	.37*** (.10)	.14* (.05, .28)
Study 2	.35** (.14)	.43*** (.09)	.15* (.04, .32)
Eudaimonic Motivations			
Study 1	.38*** (.12)	.15* (.06)	.06* (.01, .13)
Study 2	.35** (.14)	.36*** (.07)	.13* (.03, .25)
Hedonic Motivations			
Study 1	.38*** (.12)	$.02^{ns}$ (.06)	$.01^{ns}$ (-.03, .06)
Study 2	.35** (.14)	.24*** (.06)	.09* (.02, .18)
Health Motivations			
Study 1	.38*** (.12)	$.03^{ns}$ (.07)	$.01^{ns}$ (-.04, .08)
Study 2	.35** (.14)	.33*** (.07)	.12* (.03, .23)
Explicit Islamophobia	.35** (.14)	$-.05^{ns}$ (.07)	$-.02^{ns}$ (-.09, .03)
Implicit Attitudes toward Arab Muslims	.35** (.14)	$-.04^{ns}$ (.03)	$-.03^{ns}$ (-.11, .05)

Note. ^{ns} nonsignificant, *$p \leq .05$. **$p \leq .01$. ***$p \leq .001$.

mediation analyses for Study 2 suggest a significant indirect effect (see Table 1). The inconsistent evidence related to ISIS Destruction may indicate that the findings of Study 1 were a Type II error, or the findings of Study 2 were a Type I error.

With regard to increased U.S. interventions (H4), the two studies indicate a consistent pattern. In both studies, anger-disgust positively predicted a desire for increased military and humanitarian interventions, and the significant indirect effects were consistent with mediation. These findings suggest that the elicitation of anger-disgust by graphic violence can foster desires for U.S. interventions designed to avert terrorist activities and to help victims (see Table 1).

With regard to motivations (H5), the patterns were consistent between studies for eudaimonic motivations. Anger-disgust positively predicted eudaimonic motivations and mediated the effect of condition on eudaimonic motivations. These results are consistent with Bartsch et al. (2016)—which suggest that violence and gore can motivate meaning seeking in audiences—and implicate moral emotions as the explanatory mechanism. Moreover, they reflect the non-monolithic nature of media violence and its effects (see Zillmann, 1998). With regard to hedonic and health motivations, although Study 1 did not find sig-nificant relationships from anger-disgust, Study 2 did (see Table 1). Again, this inconsistency suggests that the results of Study 1 or the results of Study 2 may have been a Type II or Type I error, respectively.

Finally, with regard to attitudes toward Arab Muslims (H6a and H6b), there was no relationship between anger-disgust and either the explicit or implicit measures. These findings indicate that higher levels of graphic violence did not lead to increased antipathy toward Arab Muslims.

DISCUSSION

Overall, the findings of the current studies suggest that displaying graphic violence—as compared to sanitizing graphic violence—elicits higher levels of moral emotions, prompting (a) greater moral sensitivity, (b) greater willingness to stop terrorist organizations, and (c) eudaimonic motivations. These effects were consistent across both studies and challenge common wisdom that sanitizing graphic violence in news coverage will lead to equivalent effects on viewers as displaying it. At the same time, there was no evidence of increased prejudice toward Arab Muslims. Although this is only one potential antisocial effect, the fact that we observed prosocial effects (e.g., greater moral sensitivity) absent antisocial effects suggests that editorial policies governing the display of graphic violence should take a more balanced approach to the issue.

Graphic Violence as Moral Motivator

If graphic violence in news coverage can help increase moral responses, then the sanitization of graphic images by news media may have costs not currently considered in the policy statements of professional journalism organizations. For example, research on genocide prevention suggests that it is only possible to prevent or quell genocides "if there is the will to act *early* and *effectively* [emphasis added]" (Totten, 2004, p. 229). By sanitizing visual evidence of atrocities, media could fail to cultivate such a will. At the same time, it is important to note that interventions designed to quell the violence could exacerbate rather than reduce the problem. Media's coverage of terrorist groups could magnify the public's concerns beyond what might be considered rational, given objective risk probabilities. This type of emotion-induced risk exaggeration is well-documented (see Lerner, Gonzalez, Small, & Fischhoff, 2003). In fact, President Obama, in his 2016 State of the Union Address, attempted to downplay the threat of ISIS to realign perceptions of the risks associated with ISIS with the actual risks posed by the group (see Grizzard, 2016). Thus, the display of graphic violence within news media is likely to be a double-edged sword; there are likely pros and cons associated with both display of graphic violence and its sanitization. Current journalism standards suggest that graphic violence should be avoided and fail to acknowledge any potential benefits associated with its display (Pumarlo, n.d.). Future research should explore the costs and benefits associated with graphic violence in a balanced way so that policy statements are supported by empirical evidence.

Group Biases and Emotions Toward Perpetrators and Victims of Graphic Violence

Viewing graphic violence is likely to lead to two effects related to social identity in intergroup conflicts (see Tajfel & Turner, 1979). The first effect is sympathy toward the victim and the second is antipathy toward the perpetrator, moderated by dispositional affiliation. This moderation can reverse the typical reflexive pattern of antipathy toward perpetrator and sympathy toward victim, whereby an ingroup member attacking an outgroup member is perceived as a positive event (see Zillmann, 2006). Although higher levels of graphic violence might increase moral sensitivity, it is also possible for this type of content to lead to negative consequences, including increased antipathy toward the perpetrators' culture. These processes suggest that displaying graphic violence might result in greater ingroup-outgroup bias, a predictor of xenophobia and racism (Brewer, 1999; Fiske, 2000). For example, when the perpetrator of violence is an outgroup member and the victim is an ingroup member, providing graphic images of the violence may serve to exacerbate ingroup bias—a situation perhaps best reflected in anti-Islamic sentiment following the attacks of September 11, 2001 (Kaplan, 2006).

Conversely, when the perpetrator of violence is from the ingroup, displaying graphic consequences might serve to correct for ingroup bias, especially when the victims are outgroup members. Viewing the actual consequences of graphic violence committed by one's ingroup against an outgroup should increase sympathy toward the outgroup while decreasing identification with one's ingroup. This logic would explain why German civilians were forced to tour Nazi concentration camps. In fact, the documentary *Death Mills* (Wilder & Burger, 1945) showed that many Germans had strong physical and emotional reactions to seeing the horrors of the death camps. Thus, exposing individuals to graphic violence committed by their ingroup may be a method to reduce ingroup bias and rehumanize a dehumanized outgroup.

With regard to outgroup perpetrator–outgroup victim, depictions of graphic violence might serve to balance increased antipathy toward the perpetrators against increased sympathy toward the victims. Notably, this is the situation present in our stimuli (both the perpetrators and victims were Arab), and this explanation is consistent with our findings. This balancing should lead to no change in prejudice toward the outgroup but may potentially lead to increases in sympathy. Seeing the consequences of violence committed by ISIS on other Arab Muslims may allow viewers to characterize ISIS not as an outgroup threatening the ingroup, but rather as terrorists whose victims are primarily Arab Muslims (see Ibrahim, 2015).

Finally, it is important to point out that the ingroup–outgroup distinction as just discussed is somewhat fluid (Brewer, 1999). For example, when playing in the World Cup, the French could view Belgians as being members of an outgroup; when both are victims of terrorist attacks, they may view each other as part of the same ingroup. Thus, the potential effects just discussed must be considered in conjunction with the fluidity of ingroup–outgroup status.

Limitations

The current study suffers from some limitations. The first consists of the potential ceiling effects observed in the anger-disgust measure. Despite attempts to reduce ceiling effects in the current studies (e.g., using the anchors *extremely* and *not at all*), many participants scored at the max of the scale. At the same time, it is important to note that this limitation should have reduced our ability to detect effects and it is entirely consistent with our hypotheses; higher levels of graphic violence served to polarize feelings of anger and disgust. Still, future research should develop and utilize more sensitive measures that are less susceptible to ceiling effects. Second, emotional responses are inherently transient. Although the current findings suggest that the elicitation of moral emotions by graphic violence can have an indirect influence on moral sensitivity, it is unclear whether these effects would be lasting. However, the stronger effects observed

45

with graphic violence should have a greater potential for persistence than weaker effects observed with sanitized violence (see Fahmy et al., 2014). Third, as a measure of implicit bias, the IAT is not without controversy; critics argue that it lacks diagnostic utility and it is no better at predicting prejudiced behaviors than explicit measures (see Oswald et al., 2013). Although our results are similar for both the explicit and implicit measures, if the IAT lacks diagnostic utility, there may have been undetected increases in implicit prejudice toward Arab Muslims (i.e., a Type II error). Fourth, the current study utilized a convenience sample. Thus, it would be useful to replicate in a larger, nationally representative sample. A final limitation relates to graphic violence and selective exposure (see McKinley & Fahmy, 2011). One reason news media sanitize graphic violence is fear of alienating viewers and reducing audience size (Keith et al., 2006). If graphic violence exerts a positive influence on moral sensitivity while exerting a negative influence on audience size, gains in public concern could be negated by decreased viewership. Future research should examine these competing processes.

CONCLUSION

Our findings suggest that graphic violence in news can act as a moral motivator. Humans today—especially in Western countries—are more insulated from acts of violence than at any other point in history (Pinker, 2011). This insulation may lead to a lack of attention to the suffering of others, especially if clear evidence of that suffering is not overt. News media could combat this insulation by displaying, in graphic ways, the consequences associated with humanitarian crises. Although this conclusion is likely to rankle critics, the results of the current studies suggest that graphic violence may yield at least some positive effects. Perhaps it is only through confronting tragedy and horror head on that we become motivated to put an end to it.

REFERENCES

Agerström, J., & Rooth, D. O. (2009). Implicit prejudice and ethnic minorities: Arab-Muslims in Sweden. *International Journal of Manpower, 30*, 43–55. doi:10.1108/01437720910948384

Arab-Muslim IAT (Version 03-13-2015) [Inquisit Script]. Retrieved from http://www.millisecond. com/download/library/ArabMuslimIAT/

Aust, C. F., & Zillmann, D. (1996). Effects of victim exemplification in television news on viewer perception of social issues. *Journalism & Mass Communication Quarterly, 73*, 787–803. doi:10.1177/107769909607300403

Bartsch, A., & Mares, M. L. (2014). Making sense of violence: Perceived meaningfulness as a predictor of audience interest in violent media content. *Journal of Communication, 64*, 956–976. doi:10.1111/jcom.12112

Bartsch, A., Mares, M. L., Scherr, S., Kloß, A., Keppeler, J., & Posthumus, L. (2016). More than shoot-em-up and torture porn: Reflective appropriation and meaning-making of violent media content. *Journal of Communication, 66*, 741–765. doi:10.1111/jcom.12248

Borland, R., Wilson, N., Fong, G. T., Hammond, D., Cummings, K. M., Yong, H. H., . . . McNeill, A. (2009). Impact of graphic and text warnings on cigarette packs: Findings from four countries over five years. *Tobacco Control, 18*, 358–364. doi:10.1136/tc.2008.028043

Brewer, M. B. (1999). The psychology of prejudice: Ingroup love and outgroup hate? *Journal of Social Issues, 55*, 429–444. doi:10.1111/0022-4537.00126

Carr, D. (2014, September 7). *With videos of killings, ISIS sends medieval message by modern method. The New York Times.* Retrieved from http://goo.gl/SjlRyk

Cook, B. (2001). Over my dead body: The ideological use of dead bodies in network news coverage of Vietnam. *Quarterly Review of Film & Video, 18*, 203–216. doi:10.1080/10509200109361524

Das, E., Bushman, B. J., Bezemer, M. D., Kerkhof, P., & Vermeulen, I. E. (2009). How terrorism news reports increase prejudice against outgroups: A terror management account. *Journal of Experimental Social Psychology, 45*, 453–459. doi:10.1016/j.jesp.2008.12.001

Eden, A., Tamborini, R., Grizzard, M., Lewis, R., Weber, R., & Prabhu, S. (2014). Repeated exposure to narrative entertainment and the salience of moral intuitions. *Journal of Communication, 64*, 501–520. doi:10.1111/jcom.12098

Elson, M., & Ferguson, C. J. (2014). Twenty-five years of research on violence in digital games and aggression: Empirical evidence, perspectives, and a debate gone astray. *European Psychologist, 19*, 33–46. doi:10.1027/1016-9040/a000147

Fahmy, S. (2005). Photojournalists' and photo editors' attitudes and perceptions: The visual coverage of 9/11 and the Afghan War. *Visual Communication Quarterly, 12*, 146–163. doi:10.1080/15551393.2005.9687454

Fahmy, S., Bock, M. A., & Wanta, W. (2014). *Visual communication theory and research: A mass communication perspective.* New York, NY: Palgrave Macmillan.

Ferguson, C. J. (2010). Blazing angels or resident evil? Can violent video games be a force for good? *Review of General Psychology, 14*, 68–81. doi:10.1037/a0018941

Ferguson, C. J. (2015). Does media violence predict societal violence? It depends on what you look at and when. *Journal of Communication, 65*, E1–E22. doi:10.1111/jcom.12129

Fiske, S. T. (2000). Stereotyping, prejudice, and discrimination at the seam between the centuries: Evolution, culture, mind, and brain. *European Journal of Social Psychology, 30*, 299–322. doi:10.1002/(SICI)1099-0992(200005/06)30:3<299::AID-EJSP2>3.0.CO;2-F

Graham, J., Nosek, B. A., Haidt, J., Iyer, R., Koleva, S., & Ditto, P. H. (2011). Mapping the moral domain. *Journal of Personality and Social Psychology, 101*, 366–385. doi:10.1037/a0021847

Greenwald, A. G., Nosek, B. A., & Banaji, M. R. (2003). Understanding and using the Implicit Association Test: I. An improved scoring algorithm. *Journal of Personality and Social Psychology, 85*, 197–216. doi:10.1037/0022-3514.85.2.197

Grizzard, M. (2016). The psychology behind ISIL's media usage. *HDIAC Journal, 3*, 24–29. Retrieved from https://www.hdiac.org/node/3860

Grizzard, M., Tamborini, R., Lewis, R. J., Wang, L., & Prabhu, S. (2014). Being bad in a video game can make us morally sensitive. *Cyberpsychology, Behavior, and Social Networking, 17*, 499–504. doi:10.1089/cyber.2013.0658

Haidt, J. (2003). The moral emotions. In R. J. Davidson, K. R. Scherer, & H. H. Goldsmith (Eds.), *Handbook of affective sciences* (pp. 852–870). New York, NY: Oxford University Press.

Haidt, J., & Joseph, C. (2008). The moral mind: How 5 sets of innate intuitions guide the development of many culture-specific virtues, and perhaps even modules. In P. Carruthers, S. Laurence, & S. Stich (Eds.), *The innate mind* (Vol. 3, pp. 367–391). New York, NY: Oxford University Press.

Hayes, A. F. (2013). *Introduction to mediation, moderation, and conditional process analysis: A regression-based approach*. New York, NY: Guilford Press.

Hoffner, C., Buchanan, M., Anderson, J. D., Hubbs, L. A., Kamigaki, S. K., Kowalczyk, L., ... Silberg, K. J. (1999). Support for censorship of television violence: The role of the third-person effect and news exposure. *Communication Research, 26*, 726–742. doi:10.1177/009365099026006004

Hutcherson, C. A., & Gross, J. J. (2011). The moral emotions: A social–functionalist account of anger, disgust, and contempt. *Journal of Personality and Social Psychology, 100*, 719–737. doi:10.1037/a0022408

Ibrahim, B. (2015). But ISIS kills more Muslim than non-Muslims! Retrieved from http://goo.gl/fO9WYR

Inquisit 4 [Computer software]. (2015). Seattle, WA: Millisecond Software.

Joeckel, S., Bowman, N. D., & Dogruel, L. (2012). Gut or game? The influence of moral intuitions on decisions in video games. *Media Psychology, 15*, 460–485. doi:10.1080/15213269.2012.727218

Kaplan, J. (2006). Islamophobia in America?: September 11 and Islamophobic hate crime. *Terrorism and Political Violence, 18*, 1–33. doi:10.1080/09546550500383209

Keith, S., Schwalbe, C. B., & Silcock, B. W. (2006). Images in ethics codes in an era of violence and tragedy. *Journal of Mass Media Ethics, 21*, 245–264. doi:10.1207/s15327728jmme2104_3

Krämer, N., Eimler, S. C., Neubaum, G., Winter, S., Rösner, L., & Oliver, M. B. (2016). Broadcasting one world: How watching online videos can elicit elevation and reduce stereotypes. *New Media & Society*. Advance online publication. doi:10.1177/1461444816639963

Kratzer, R. M., & Kratzer, B. (2003). How newspapers decided to run disturbing 9/11 photos. *Newspaper Research Journal, 24*, 34–47. doi:10.1177/073953290302400104

Lee, S. A., Gibbons, J. A., Thompson, J. M., & Timani, H. S. (2009). The Islamophobia scale: Instrument development and initial validation. *The International Journal for the Psychology of Religion, 19*, 92–105. doi:10.1080/10508610802711137

Lerner, J. S., Gonzalez, R. M., Small, D. A., & Fischhoff, B. (2003). Effects of fear and anger on perceived risks of terrorism a national field experiment. *Psychological Science, 14*, 144–150. doi:10.1111/1467-9280.01433

Lichtblau, B. (2015, December 17). Crimes against Muslim Americans and mosques rise sharply. *The New York Times*. Retrieved from http://goo.gl/r8S4mn

Markey, P. M., Males, M. A., French, J. E., & Markey, C. N. (2015). Lessons from Markey et al. (2015) and Bushman et al. (2015): Sensationalism and integrity in media research. *Human Communication Research, 41*, 184–203. doi:10.1111/hcre.12057

McKinley, C. J., & Fahmy, S. (2011). Passing the "breakfast test": Exploring the effects of varying degrees of graphicness of war photography in the new media environment. *Visual Communication Quarterly, 18*, 70–83. doi:10.1080/15551393.2011.574060

Oliver, M. B., Hartmann, T., & Woolley, J. K. (2012). Elevation in response to entertainment portrayals of moral virtue. *Human Communication Research, 38*, 360–378. doi:10.1111/j.1468-2958.2012.01427.x

Oswald, F. L., Mitchell, G., Blanton, H., Jaccard, J., & Tetlock, P. E. (2013). Predicting ethnic and racial discrimination: A meta-analysis of IAT criterion studies. *Attitudes and Social Cognition, 105*, 171–192. doi:10.1037/a0032734

Park, J., Felix, K., & Lee, G. (2007). Implicit attitudes toward Arab-Muslims and the moderating effects of social information. *Basic and Applied Social Psychology, 29*, 35–45. doi:10.1080/01973530701330942

Pinker, S. (2011). *The better angels of our nature: A history of violence and humanity*. London, UK: Viking Penguin.

Prinz, J., & Nichols, S. (2010). Moral emotions. In J. M. Doris (Ed.), *The moral psychology handbook* (pp. 111–146). New York, NY: Oxford University Press.

PROCESS (Version 2.16) [Computer Software]. Retrieved from http://processmacro.org/

Pumarlo, J. (n.d.). *SPJ Ethics committee position papers: Reporting on grief, tragedy and victims*. Retrieved from http://www.spj.org/ethics-papers-grief.asp

Scharrer, E., & Blackburn, G. (2015). Images of injury: Graphic news visuals' effects on attitudes toward the use of unmanned drones. *Mass Communication and Society, 18*, 799–820. doi:10.1080/15205436.2015.1045299

Tajfel, H., & Turner, J. (1979). An integrative theory of inter-group conflict. In J. A. Williams, & S. Worchel (Eds.), *The social psychology of inter-group relations* (pp. 33–47). Belmont, CA: Wadsworth.

Tamborini, R. (2013). A model of intuitive morality and exemplars. In R. Tamborini (Ed.), *Media and the moral mind* (pp. 43–74). New York, NY: Routledge.

Tamborini, R., Prabhu, S., Lewis, R. J., Grizzard, M., & Eden, A. (2016). The influence of media exposure on the accessibility of moral intuitions and associated affect. *Journal of Media Psychology*, Advance online publication. 1–12. doi:10.1027/1864-1105/a000183

Tangney, J. P., Stuewig, J., & Mashek, D. J. (2007). Moral emotions and moral behavior. *Annual Review of Psychology, 58*, 345–372. doi:10.1146/annurev.psych.56.091103.070145

Totten, S. (2004). The intervention and prevention of genocide: Sisyphean or doable? *Journal of Genocide Research, 6*, 229–247. doi:10.1080/1462352042000225967

Van Der Molen, J. H. W. (2004). Violence and suffering in television news: Toward a broader conception of harmful television content for children. *Pediatrics, 113*, 1771–1775.

Vettehen, P. H., Nuijten, K., & Beentjes, J. (2005). News in an age of competition: The case of sensationalism in Dutch television news, 1995-2001. *Journal of Broadcasting & Electronic Media, 49*, 282–295. doi:10.1207/s15506878jobem4903_2

Wilder, B., & Burger, H. (Directors). (1945). *Death mills* [Motion picture]. United States: U.S. Army Signal Corps.

Zillmann, D. (1998). The psychology and appeal of portrayals of violence. In J. Goldstein (Ed.), *Why we watch: The attractions of violent entertainment*. New York, NY: Oxford University Press.

Zillmann, D. (2002). Exemplification theory of media influence. In J. Bryant & D. Zillmann (Eds.), *Media effects: Advances in theory and research* (2nd ed., pp. 19–41). Mahwah, NJ: Erlbaum.

Zillmann, D. (2006). Empathy: Affective reactivity to others' emotional experiences. In J. Bryant & P. Vorderer (Eds.), *Psychology of entertainment* (pp. 151–182). Mahwah, NJ: Erlbaum.

Zillmann, D., & Gan, S. (1996). Effects of threatening images in news programs on the perception of risk to others and self. *Medienpsychologie: Zeitschrift Für Individual- Und Massenkommunikation, 8*, 317–318.

Online Surveillance's Effect on Support for Other Extraordinary Measures to Prevent Terrorism

Elizabeth Stoycheff, Kunto A. Wibowo, Juan Liu, and Kai Xu

The U.S. National Security Agency argues that online mass surveillance has played a pivotal role in preventing acts of terrorism on U.S. soil since 9/11. But journalists and academics have decried the practice, arguing that the implementation of such extraordinary provisions may lead to a slippery slope. As the first study to investigate empirically the relationship between online surveillance and support for other extraordinary measures to prevent terrorism, we find that perceptions of government monitoring lead to increased support for hawkish foreign policy through value-conflict associations in memory that prompt a suppression of others' online and offline civil liberties, including rights to free speech and a fair trial. Implications for the privacy–security debate are discussed.

Elizabeth Stoycheff (Ph.D., The Ohio State University, 2013) is an assistant professor in the Department of Communication at Wayne State University. Her research interests include the role of digital technologies in democratic development and sustainability.

Kunto A. Wibowo (M.A., The Hague University, 2009) is a lecturer at Universitas Padjadjaran in Indonesia. His research interests include the consequences of algorithms and big data on political attitudes and behaviors.

Juan Liu (Ph.D., Wayne State University, 2017) is an assistant professor at Columbus State University. Her research interests include social media, digital divides, online expression, and political polarization.

Kai Xu (M.A., Western Kentucky University/Wayne State University, 2012/2015) is a doctoral student in the Department of Communication at Wayne State University. His research interests include media framing and misinformation in authoritarian contexts.

Color versions of one or more of the figures in the article can be found online at www.tandfonline.com/hmcs.

Mass online surveillance has been justified as a primary tool in the United States' continued fight against terror. But independent investigation has shown that it has had "no discernable impact on preventing acts of terrorism" (Bergen, Sterman, Schneider, & Cahall, 2014, p. 2), and initial research suggests that surveillance poses spillover effects on two central tenets of democracy: individuals' access to information and freedom of speech (Penney, 2016; Rainie & Madden, 2015; Stoycheff, 2016). This study uniquely uncovers two others: political tolerance and hawkish policy attitudes.

Omnipresent government surveillance (Bernal, 2016) serves as a continued reminder of the threat of terrorism and highlights how average Americans have sacrificed online civil liberties for national security. Previous research suggests that when individuals are primed of an existential threat—and the specific costs associated with protecting it (Peffley, Knigge, & Hurwitz, 2001)—they may become less tolerant toward individuals and groups suspected to be involved (Davis & Silver, 2004; Saleem, Prot, Anderson, & Lemieux, 2015) and more supportive of hawkish military actions (Hirschberger et al., 2009; Pyszczynski et al., 2006) to eradicate it.

Using an online experiment that primes individuals of government surveillance, we advance and test a serial mediation model that examines how perceptions of online monitoring increases support for three other extraordinary measures in the fight against terror. We argue that perceptions of online surveillance trigger value-conflict associations in memory that prompt support for restricting others' online (e.g., website censorship) and offline (e.g., unlawful detainment) civil liberties for groups and individuals associated with terrorism. This intolerance, in turn, feeds support for hawkish foreign policies that advance military intervention abroad in the name of terrorism prevention. These findings contribute to the small, but growing, body of literature on surveillance's potential effects. We then situate these results in the context of the ongoing privacy–security debate.

TERRORISM AS A SECURITIZED ISSUE

Since 9/11, terrorism has become a mainstay of the United States' domestic and foreign policy (Anderson, 2015; Bapat, 2011; Choi & James, 2016), preoccupying federal funding (Bandyk, 2010), and media coverage (Aday, 2010; Bowman, Lewis, & Tamborini, 2014; Fahmy, 2005, 2010; Friedman, 2008; Gadarian, 2014; Haridakis & Rubin, 2005; Houston, 2009; Lewis & Reese, 2009; Morin, 2016; Reynolds & Barnett, 2003; Ross & Bantimaroudis, 2006; Scheufele, Nisbet, & Ostman, 2005; Yarchi, Wolfsfeld, Sheafer, & Shenhav, 2013). Political elites from both parties and the media have framed the issue as an existential threat to the physical, cultural, and social ways of life in America, through a theoretical process known as securitization (Vultee, 2010). Securitization occurs through repeated association of an issue as a threat to the United States' very existence—including its sovereignty, territorial integrity, or safety of a large number of Americans (Jacobson, 2016)—often facilitated by the media (Vultee, Lukacovic, & Stouffer, 2015). When effective, securitized issues suspend the normal rules of order, such that extraordinary measures are deemed necessary and are taken to combat them (Buzan, Waever, & De Wilde, 1998).

Mediated discourse surrounding terrorism is laden with examples of its securitized status, despite that it remains too rare an occurrence to be classified as such (Mueller, 2006; Mueller & Stewart, 2015). Several of the 2016 presidential nominees on the campaign trail declared ISIS and the "global jihadist movement" as threats to U.S. existence (Jacobson, 2016). In particular, a 2016 report from a well-known think-tank ominously argued that terrorist organizations currently "accelerate the collapse of world order … and endanger American values and way of life" (Kagan, Kagan, Cafarella, Gambhir, & Zimmerman, 2016). Leadership at the FBI has even securitized cyberterrorism, contending that it can wipe out cities and produce casualties (Henry, 2011).

In response, the U.S. government—under both Democratic and Republican leadership—has adopted a number of extraordinary provisions in the fight against terrorism, including the U.S. war in Afghanistan, which encompassed aerial bombing campaigns and ground troop invasions, economic sanctions on states thought to harbor terrorists, and ongoing operations at the Guantanamo Bay detention camp. But one of the most surreptitious and understudied deviations has been the widespread use of online mass surveillance programs (Bauman et al., 2014; Greenwald, 2014; Lischka, 2017; Lyon, 2014).

MASS ONLINE SURVEILLANCE

To combat the threat of terrorism, the U.S. government has engaged in three distinct layers of online mass surveillance (Lyon, 2015): (a) upstream programs that provide the National Security Agency with access to optical cables that carry information between the United States and foreign countries; (b) the use of malware to install spyware on individuals' personal devices; and (c) bulk collection and storage of online data, known as PRISM, which clandestinely obtains content and metadata from the servers of widely used Internet companies, including Apple, Facebook, YouTube, Skype, Yahoo!, and Google, among others. Originally thought to target only foreign nationals, reporting over the past several years has revealed that the PRISM program is far less discriminant, requiring only 51% confidence of an individual's identity (Lee, 2013), which consequently has cast a dragnet over the daily communications of ordinary, law-abiding Americans (Greenwald, 2014; Lyon, 2014). In addition, collected and stored data are shared widely among other governmental agencies to be used for purposes other than terrorism prevention (Balko, 2016). In 2013, the President's Review Group on Intelligence and Communication Technologies (Clarke, Morell, Stone, Sunstein, & Swire, 2013) pushed back against these policies and recommended a complete end to storage of privately held data obtained by the U. S. government, cautioning that "excessive surveillance and unjustified secrecy can threaten the civil liberties, public trust, and the core processes of self-government" (p. 12). To date, these recommendations have not been incorporated into law.

In light of these revelations, the vast majority of Americans report awareness of governmental programs that may monitor their online activities (Pew Research, 2015), and Twitter opinion leaders on this issue have decried their use (Reddick, Chatfield, & Jaramillo, 2015). Two of the first empirical studies examining individuals' perceptions of online mass surveillance have shown that it has the potential to deeply undermine the democratic process by stifling individuals' online political information seeking behaviors and political discussions (Penney, 2016; Stoycheff, 2016; see also Rainie & Madden, 2015). Specifically, Penney's (2016) systematic analysis showed a 25% drop in the views of sensitive Wikipedia articles after the National Security Agency's surveillance programs gained widespread media attention in 2013. Stoycheff (2016) documented a conditional effect of perceptions of surveillance on individuals' willingness to discuss sensitive political issues on Facebook, a result substantiated by survey data that show how individuals believe this to be an important issue but refuse to talk about it in online contexts (Rainie & Madden, 2015).

VALUE CONFLICT AND OTHER EXTRAORDINARY MEASURES

Artifacts of online surveillance, including terms of use, content agreements, news stories, and privacy warnings, abound in individuals' daily online lives, serving as continual reminders of the threat of terrorism and its securitized status. Although Americans do value digital privacy, they also strive to protect themselves against threats like terrorism (Rainie & Maniam, 2016), creating a continuous tension between balancing one's civil liberties and one's security. When a government imposes extraordinary measures—like surveillance—to combat a threat, the balance is disrupted, prioritizing security over individual liberties. We anticipate that this prioritization will be activated upon exposure to surveillance primes and bring to mind other value-conflict associations in memory that can be used in forming subsequent judgments, including those about foreign policy initiatives. There is a commanding body of literature that shows how threats to one's security evoke support for aggressive military action (Hetherington & Suhay, 2011) even years after the immediate threat has subsided (Gadarian, 2010). For example, Huddy and colleagues (2005) found that perceptions of an external threat increase both the public's support for retaliatory military action and approval of the government that oversees it.

Moreover, we expect that this effect can be observed indirectly through a reduction in tolerance as individuals engage in complex political reasoning that also prioritizes security over civil liberties. Political tolerance is necessary to ensure a free-flowing marketplace of information and the right of all citizens to participate in politics. But the question of how far citizens are willing to extend basic civil liberties to nonconformist groups (Stouffer, 1955) and those with objectionable political ideas (Sullivan, Piereson, & Marcus, 1982) has long been debated. Conflicting values have been shown to make otherwise tolerant individuals significantly less open-minded when confronted with the understanding that they must shoulder some of the "costs" associated with protecting others (Peffley et al., 2001). In the fight against terrorism, mass surveillance indeed serves as a reminder that security has "cost" average citizens some digital privacy, leading us to expect that individuals made aware of surveillance will similarly resolve this value conflict by prioritizing their own security over defending others' online civil liberties.

Although rights associated with online surveillance are largely confined to the digital context, the threat of terrorism is not. Once an individual has adopted a security-over-liberty position, it is likely to be readily accessible in memory for use in evaluations of other extraordinary measures in the fight against terrorism, including the protection of individuals' offline civil liberties and military action abroad.

Many communication scholars have examined how various news media can prime citizens to the existential threat of terrorism, which in turn reduces support for extending civil liberties to unpopular groups (Saleem et al., 2015; Scheufele et al., 2005; White, Duck, & Newcombe, 2012) and influences foreign policy attitudes (Gadarian, 2010). Here, we advance a serial mediation model that explains the process through which perceptions of government surveillance should operate as a similar prime, provoking an effortful, memory-based processing that contributes to support for other extraordinary measures through value-conflict prioritizations. Specifically, when individuals perceive their online activities are susceptible to government surveillance, mental nodes that pertain to the government's prioritization of national security over digital privacy rights will be activated and made available for use in short-term memory (Domke, Shah, &

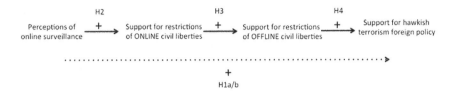

FIGURE 1 Hypothesized serial mediation model.

Wackman, 1998; Hastie & Park, 1986; Roskos-Ewoldsen, Roskos-Ewoldsen, & Dillman Carpenter, 2009). These activated memory nodes will then be over-sampled when individuals are asked to form their own judgments about other extraordinary measures in the fight against terrorism. Thus, we first anticipate that *perceptions of online surveillance will increase support for hawkish foreign policy attitudes* (H1a).

Once this security-over-liberty position has been made accessible, we expect it will trigger a chain reaction through tolerance judgments, such that there will be evidence of an *indirect effect of surveillance on support for hawkish foreign policy attitudes* (H1b). This indirect effect is expected to occur by first weighing the personal costs associated with surveillance (e.g., one's online rights) with the costs of protecting others' digital rights, such that *perceptions of online surveil-lance will be positively associated with support for online civil liberty restric-tions* (H2). Because the threat of terrorism extends to the offline world, we anticipate that value-conflict associations previously activated will again be oversampled, such that *support for online civil liberty restrictions will be posi-tively associated with support for offline civil liberty restrictions* (H3). Last, consistent with previous literature (e.g., Gadarian, 2010; Hetherington & Suhay, 2011; Huddy & Feldman, 2011), political intolerance has been estab-lished as a consistent predictor of hawkish foreign policy attitudes, which we expect to be replicated here, such that *support for restrictions of offline civil liberties will contribute to hawkish foreign policy attitudes to prevent terrorism* (H4). The proposed serial model and corresponding hypotheses are depicted in Figure 1.

METHODOLOGY

Procedure and Sample

To investigate the serial mediation just proposed, we grounded a study in ecologically valid political events. An online experiment was embedded in a Qualtrics questionnaire in August 2014, during a period when Congress had introduced versions of the USA Freedom Act, a law that sought to end bulk collection of Americans' phone records, a sister program to PRISM that was also exposed during the 2013 National Security Agency classified-document leak. The USA Freedom Act ultimately passed with overwhelming support in mid-2015, but mass online surveillance has yet to be regulated. During this same period, President Obama authorized U.S. airstrikes against terrorist targets in Iraq, which have since been extended to Syria.

The study's participants ($N = 152$) were recruited via Amazon's Mechanical Turk (MTurk), a voluntary, online worker pool that reflects this study's target population of American internet users. Individuals needed to reside in the U.S., have their work approved at least 95% of the time (MTurk approval rate), and be at least 18 years old to be eligible to participate. Although MTurk relies on a convenience sample—which is permissible for investigating communication processes—its makeup was comparable to the U.S. Internet population's age and gender distributions.

Measures

The primary independent variable, *perceptions of surveillance*, was measured dichotomously and manipulated by the researchers. Specifically, we altered the experiment's introductory screen that accompanied our university's institutional logo and Institutional Review Board–required online privacy statement remind-ing participants that their online communication was susceptible to interception from third parties. A randomly assigned subset of participants ($n = 60$) were exposed to an additional National Security Agency logo and statement that specified the U.S. government could be one of these third parties, using the following language: "Activity on this site may be monitored by the U.S. National Security Agency." This manipulation, displayed in the appendix, attempted to emulate websites' terms of agreement statements that notify users their data can and may be used without their consent and closely mirrors the stimuli employed in Stoycheff's (2016) surveillance chilling study. The remaining subset of parti-cipants encountered only the mandatory Institutional Review Board privacy statement. This variable was quantified dichotomously, with the surveillance prime coded high. This stimuli was intended to activate beliefs that surveillance was an extraordinary measure in the fight against terrorism, and a dichotomous manipulation check demonstrated that the surveillance prime did indeed prompt 67% of surveilled individuals to consider government monitoring to be of the ordinary or "unusual," compared with only 8% of those not exposed to the prime, $\chi^2 = 17.75$, $p < .001$ (sample size = 152; $df = 1$).

The study's two mediating variables, *support for restricting online civil liberties* and *support for restricting offline civil liberties*, were adapted from Crowson and DeBacker (2008) and assessed on 7-point Likert scales, such that higher scores were indicative of greater civil liberty restriction. Online intolerance measured participants' level of agreement that the U.S. government should monitor the emails of all foreign-born nationals, censor e-mails that contain suspected terrorist messages, and censor websites that are affiliated with terrorist organizations (M = 4.62, SD = 1.98; Cronbach's α = 0.83). Offline intolerance, in the fight against terrorism, similarly captured participants' support for the U.S. government to hold suspected individuals indefinitely without trial, monitor individuals' financial transactions, try suspected persons with military tribunals, and generally allow the use of racial profiling (M = 4.45, SD = 1.84; Cronbach's α = 0.82). Our dependent variable, *hawkish terrorism foreign policy*, used an 11-point item that asked the degree to which participants supported or opposed the continuation of air strikes against terrorist targets abroad, with higher values indicating greater support (M = 7.01, SD = 2.93).

The sample had an average age of 34.75 (SD = 11.20), was 49% female, and identified as slightly Democratic on a 1–7 scale ranging from *strong Democrat* to *strong Republican* (M = 3.30, SD = 1.58). Four participants identified a preference for another political party and were excluded listwise. As expected from previous research (Crowson & DeBacker, 2008; Huddy et al., 2005), one's political party preference positively correlated with the study's dependent and mediating variables, such that Republicans expressed greater support for hawkish policy preferences (Pearson's r = .16, p < .05) and restrictions of individuals' online (Pearson's r = .19, p < .05) and offline (Pearson's r = .30, p < .01) civil liberties. Racially, the sample was comprised of 68.4% who identified as Caucasian, 21.7% as Asian, 5.9% as Black or African American, and 4% as another race. On seven-item scales, individuals indicated they were slightly politically interested (M = 4.74, SD = 1.63) and paid above-average attention to news about terrorism and international issues (M = 4.99, SD = 1.40, Pearson's r = .64).

RESULTS

To obtain a baseline understanding of the nature of perceived surveillance, an independent-samples t test revealed significant mean differences across experimental conditions on individuals' support for both online, $t(149)$ = –2.02, p < .05, and offline, $t(149)$ = –2.50, p < .05, civil liberty restrictions, as shown in Table 1. To test the direct and indirect effects of perceived surveillance on hawkish foreign attitudes hypothesized in H1a and H1b, we conducted ordinary least squares regression analyses using Hayes's (2013) PROCESS macro, Model 6, with 5,000 bootstrap samples. This model enables the simultaneous examination of direct and indirect effects on the dependent variable of interest, hawkish foreign policy through the use of multiple mediators. The model was fitted with the surveillance prime and support for online and offline restrictions of civil liberties.

TABLE 1
Means of Dependent Variables by Experimental Group and *t*-Test Comparison

	Surveillance M (SD)	No Surveillance M (SD)	t-Test
Online civil liberty restrictions	5.02 (1.93)	4.36 (1.98)	−2.02*
Offline civil liberty restrictions	4.91 (1.77)	4.16 (1.82)	−2.50*
Hawkish terrorism foreign policy	7.47 (3.16)	6.71 (2.76)	ns

*p < .05

Model results provide no evidence of a direct effect (H1a), but they do indicate a positive significant indirect effect of surveillance on hawkish foreign policy through the two proposed mediators ($\beta = 0.28$, $SE = 0.16$, $p < .05$), in support of H1b. The prime and mediating variables explained 18% of the variance in one's hawkish foreign policy attitudes. To further explore this indirect relationship, we examined each step in the mediated pathway. In the first mediated relationship, exposure to the surveillance prime lead to greater support for online restrictions of civil liberties ($\beta = 0.64$, $SE = 0.33$, $p < .05$), such that those primed of surveillance were more likely to restrict the digital rights of others, as hypothesized in H2.

Support for restricting online rights exhibited a positive and significant association with support for restricting offline rights ($\beta = 0.73$, $SE = 0.05$, $p < .001$), as predicted in H3. Last, individuals who supported offline restrictions were more likely to hold hawkish attitudes on terrorism policy ($\beta = 0.60$, $SE = 0.20$, $p < .01$), lending evidence for H4. Neither perceptions of surveillance nor support for restricting online civil liberties exhibited significant direct effects. The hypothesized results are visualized in Figure 2. These total indirect effects demonstrate a nuanced process through which online surveillance has the potential to curtail civil liberties and advance support for military force. Having been reminded that one is susceptible to online surveillance leads individuals to adopt a security-over-liberties prioritization that is subsequently used to form judgments about other extraordinary measures in the fight against terrorism, ultimately restricting others' online and, indirectly, offline civil liberties, which in turn is associated with a hawkish foreign policy orientation.

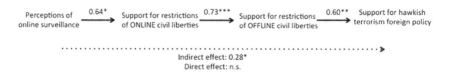

FIGURE 2 Resulting serial mediation model.

DISCUSSION

Limitations

Prior to a discussion of our findings, it's important to contextualize these results by first noting their methodological limitations. This study prioritized internal control over external validity by relying upon a nongeneralizable sample of heavy Internet users (Casler, Bickel, & Hackett, 2013). Although the study's population parameters are not representative, these users may actually demonstrate a conservative test of surveillance effects, as this group may be more mindful or sensitive about surveillance and thus more resistant to attitudinal changes after a one-time prime exposure.

In this prioritization of internal control, the manipulation allowed us to test a causal pathway between perceived surveillance and support for military intervention. Although the results indicated no direct relationship, we did uncover a theoretically and normatively meaningful indirect pathway through value-conflict associations in memory that prompted support for restricting others' civil liberties. The absence of a direct relationship should be interpreted with caution, as such effects may go undetected in instances with low measurement precision (Rucker, Preacher, Tormala, & Petty, 2011), which was indeed a significant limitation here. Our dependent variable relied on a single-item construct, and thus only tapped one dimension of individuals' support for hawkish foreign policy, albeit an integral one in the fight against terrorism.[1] Other dimensions include increasing military spending, extending resources to opposition groups, and deploying more ground troops and advisors to the region, all of which should be incorporated in future measurement. Again, this limitation likely produced a conservative test of these effects, as it taps only one dimension of a multifaceted construct. Moving forward, more robust measurement will help assess the magnitude of both direct and indirect pathways.

Implications

Despite these limitations, this study contributes to the small, but growing, body of empirical scholarship revealing that online mass surveillance does indeed pose

---

[1] Subsequent data collection with a second MTurk sample ($N = 51$) shows this single item correlates well with other existing or proposed hawkish policies in the war against terrorism, including the deployment of U.S. military advisors to the region (Pearson's $r = .57$), the deployment of U.S. ground troops to fight terrorist targets (Pearson's $r = .77$), increased military spending as a portion of the U.S. budget to combat terrorism (Pearson's $r = .60$), and increased U.S. military aid to groups that oppose and combat ISIS threats (Pearson's $r = .52$). Further, this item is significantly associated with the belief that ISIS poses an existential threat to the United States (Pearson's $r = .49$), much to the same degree as the other indicators of hawkishness just listed.

spillover effects. In addition to chilling online search behaviors and political discussion (Penney, 2016; Stoycheff, 2016), our results suggest that surveillance may contribute to a culture of mistrust and suspicion by prompting value-conflict associations in memory, as others have foreshadowed (Bernal, 2016; Campbell, 2004; O'Connor & Jahan, 2014). In particular, surveillance contributes to support for a curtailment of others' online and, indirectly, offline civil liberties. These findings complement the work of other scholars and public opinion surveys that have shown an increase in discriminatory attitudes toward Muslims (Saleem et al., 2015) and a suspicion of refugees and immigrants from Muslim countries ("Donald Trump's," 2016) as a result of terrorism's perceived existential threat. Further work that explicitly explores whether and how surveillance erodes societal or governmental trust are important next steps.

Extending the rights of free speech, association, and a fair trial to *all* citizens is a hallmark of a democratic society. Practices that jeopardize these principles undermine the very presumption of innocence, which has been troublingly reflected in the "nothing to hide" sentiments often advanced surveillance supporters (Greenwald, 2014; McStay, 2017; Mols & Janssen, 2017). These individuals contend that the government may monitor their online activities because they are not concerned with concealing any misdeeds or wrongdoing. In doing so, privacy is delegitimized as fundamental to human nature in the development of one's identity and the self.

Another common sentiment that undermines the value of civil liberties is that "the ends justify the means" (Mols & Janssen, 2017). Discussions surrounding the trade-off between privacy and security have largely been framed as an individual's right to privacy versus a collective right to existential security, which helps explain why individuals are often willing to part with their rights for national good (e.g., Huddy, Khatid, & Capelos, 2002). By relying on elite sources that rationalize surveillance as essential to terrorism prevention and public safety, media (Bernal, 2016; Lischka, 2017). Asymmetrical empirical research further aggravates this divide. In communication-related journals alone, the effects of media coverage about terrorism are well documented (e.g., Aday, 2010; Bowman et al., 2014; Fahmy, 2005, 2010; Friedman, 2008; Gadarian, 2014; Houston, 2009; Lewis & Reese, 2009; Morin, 2016; Reynolds & Barnett, 2003; Ross & Bantimaroudis, 2006; Scheufele et al., 2005; Yarchi et al., 2013), whereas empirical research that investigates the consequences of online surveillance is still in its infancy (O'Connor & Jahan, 2014; Penney, 2016; Stoycheff, 2016). This study helps lend quantifiable support that surveillance extends beyond an individual's freedom of privacy; it also impacts an array of other civil liberties, including rights to expression and a due process, which envelop entire groups and communities. In light of this evidence, we reiterate Bernal's (2016) call to recast the debate as one between the government's need to provide security and protect the freedoms of all citizens.

At each step in our serialized model, support for extraordinary measures to prevent terrorism increases in intensity. Surveilled individuals first increased their support for online censorship and monitoring of objectionable content, then shifted to support more severe offline restrictions, including unlawful detainment and imprisonment, and finally indicated greater support for militarized force. This indirect pathway—and the absence of an overall direct effect—suggests that perceptions of online surveillance only needed to trigger the first of these restrictions for the process to unfold. This process draws striking parallels to the boiling frog syndrome (Dixon, 2015), whereby the best way to boil a frog is to place it in cool water that is slowly heated so the frog is less sensitive to danger. In the same way, citizens often object to radical restrictions of others' civil liberties but are more accepting of small curtailments that occur incrementally. Thus, a *mere* violation of citizens' online privacy may lead to a slippery slope for the protection of other civil liberties and have consequences for actions abroad.

Unlike political intolerance, hawkish foreign policy attitudes do not undermine or threaten U.S. democracy. But such attitudes do provide further justification for additional extraordinary measures—like online surveillance—to eradicate the threat of terrorism. The process described in this study may prove to be iterative, resulting in a downward spiral, whereby surveillance stifles civil liberties and advances military action that further justifies the need for greater monitoring and surveillance. In short, awareness of extraordinary measures fuels greater support for more extraordinary measures. These results provide initial evidence that the slow decay of citizen rights has already begun.

Although the U.S. government maintains that online mass surveillance is an important tool in the fight against terrorism, it's necessary for scholars and policymakers alike to vet thoroughly the government's online data collection programs, such that any value conflict equally protects collective freedoms and security. Further work that quantifies how surveillance shapes individuals' online and offline behaviors will help ensure that surveillance itself does not become our next existential threat.

REFERENCES

Aday, S. (2010). Chasing the bad news: An analysis of 2005 Iraq and Afghanistan war coverage on NBC and Fox News channel. *Journal of Communication*, 60(1), 144–164.

Anderson, J. E. (2015). Terrorism, trade and public policy. *Research in Economics*, 69(2), 180–190.

Balko, R. (2016, March 10). Surprise! NSA data will soon routinely be used for domestic policing that has nothing to do with terrorism. *The Washington Post*. Retrieved from https://www.washingtonpost.com/

Bandyk, M. (2010, January 11). What airport security costs you: The federal government plans to beef up airport security, but that costs more than it might seem. *U.S. News and World Report.* Retrieved from http://money.usnews.com/money/business-economy/articles/2010/01/11/what-airport-security-costs-you

Bapat, N. (2011). Terrorism, democratization and US foreign policy. *Public Choice, 149*(3/4), 315–335.

Bauman, Z., Bigo, D., Esteves, P., Guild, E., Jabri, V., Lyon, D., & Walter, R. B. J. (2014). After Snowden: Rethinking the impact of surveillance. *International Political Sociology, 8,* 121–144.

Bergen, P., Sterman, D., Schneider, E., & Cahall, B. (2014). Do NSA's bulk surveillance programs stop terrorists? *New America Society.* Retrieved from http://pierreghz.legtux.org/streisand/autoblogs/frglobalvoicesonlineorg_0e319138ab63237c2d2aeff84b4cb506d936eab8/media/e1982452.Bergen_NAF_NSA20Surveillance_1_0.pdf

Bernal, P. (2016). Data gathering, surveillance and human rights: Recasting the debate. *Journal of Cyber Policy, 1*(2), 243–264.

Bowman, N., Lewis, R. J., & Tamborini, R. (2014). The morality of May 2, 2011: A content analysis of U.S. headlines regarding the death of Osama bin Laden. *Mass Communication & Society, 17*(5), 639–664.

Brown, I. (2014). Social media surveillance. *The International Encyclopedia of Digital Communication and Society, 1–7.*

Buzan, B., Waever, O., & De Wilde, J. (1998). *Security: A new framework for analysis.* Boulder, CO: Lynne Reiner.

Campbell, N. D. (2004). Technologies of suspicion: Coercion and compassion in post-disciplinary surveillance regimes. *Surveillance & Society, 2*(1), 78–92.

Casler, K., Bickel, L., & Hackett, E. (2013). Separate but equal? A comparison of participants and data gathered via Amazon's MTurk, social media, and face-to-face behavioral testing. *Computers in Human Behavior, 29*(6), 2156–2160.

Choi, S. W., & James, P. (2016). Why does the United States intervene abroad? Democracy, human rights violations and terrorism. *Journal of Conflict Resolution, 60*(5), 899–926.

Clarke, R. A., Morell, M. J., Stone, G. R., Sunstein, C. R., & Swire, P. (2013, December 12). *Liberty and security in a changing world: Report and recommendations of the President's review group on intelligence and communications technologies.* Retrieved from http://www.whitehouse.gov/sites/default/files/docs/2013-12-12_rg_final_report.pdf

Crowson, H. M., & DeBacker, T. K. (2008). Belief, motivational, and ideological correlates of human rights attitudes. *The Journal of Social Psychology, 148*(3), 293–310.

Davis, D. W., & Silver, B. D. (2004). Civil liberties vs. security: Public opinion in the context of the terrorist attacks on America. *American Journal of Political Science, 48*(1), 28–46.

Dixon, H. B. (2015). The end of privacy as we know it? *The Judges' Journal, 54*(2), 36–39.

Domke, D., Shah, D. V., & Wackman, D. B. (1998). Media priming effects: Accessibility, association, and activation. *International Journal of Public Opinion Research, 10,* 51–74.

Donald Trump's proposed Muslim Ban is likely illegal but… (2016, June 14).. *Newsweek.* Retrieved from http://www.newsweek.com/donald-trump-muslims-ban-terrorism-radical-islam-guns-orlando-shooting-legal-470470

Fahmy, S. (2005). Photojournalists' and photo editors' attitudes and perceptions: The visual coverage of 9/11 and the Afghan War. *Visual Communication Quarterly, 12*(3–4), 146–163.

Fahmy, S. (2010). Contrasting visual frames of our times: A framing analysis of English- and Arabic-language press coverage of war and terrorism. *International Communication Gazette, 72*(8), 695–717.

Friedman, B. (2008). Unlikely warriors: How four U.S. news sources explained female suicide bombers. *Journalism & Mass Communication Quarterly, 85*(4), 841–859.

Gadarian, S. K. (2010). The politics of threat: How terrorism news shapes foreign policy attitudes. *The Journal of Politics, 72*(2), 469–483.

Gadarian, S. K. (2014). Scary pictures: How terrorism imagery affects voter evaluations. *Political Communication, 31*(2), 282–302.

Greenwald, G. (2014). *No place to hide: Edward Snowden, the NSA, and the U.S. surveillance state.* New York, NY: Metropolitan Books.

Haridakis, P. M., & Rubin, A. M. (2005). Third-person effects in the aftermath of terrorism. *Mass Communication and Society, 8*(1), 39–59.

Hastie, R., & Park, B. (1986). The relationship between memory and judgment depends on whether the judgment task is memory-based or on-line. *Psychological Review, 93*(3), 258–268.

Hayes, A. F. (2013). *Introduction to mediation, moderation, and conditional process analysis: A regression-based approach.* New York, NY: Guilford Press.

Henry, S. (2011, October 20). Information Systems Security Association International Conference, Baltimore, Maryland. Retrieved from https://archives.fbi.gov/archives/news/speeches/responding-to-the-cyber-threat

Hetherington, M., & Suhay, E. (2011). Authoritarianism, threat, and Americans' support for the war on terror. *American Journal of Political Science, 55*(3), 546–560.

Hirschberger, G., Pyszczynski, T., & Ein-Dor, T. (2009). Vulnerability and vigilance: Threat awareness and perceived adversary intent moderate the impact of mortality salience on intergroup violence. *Personality and Social Psychology Bulletin, 35*(5), 597–607.

Houston, J. B. (2009). Media coverage of terrorism: A meta-analytic assessment of media use and posttraumatic stress. *Journalism & Mass Communication Quarterly, 86*(4), 844–861.

Huddy, L., & Feldman, S. (2011). Americans respond politically to 9/11: Understanding the impact of the terrorist attacks and their aftermath. *The American Psychologist, 66*(6), 455–467.

Huddy, L., Feldman, S., Taber, C., & Lahav, G. (2005). Threat, anxiety, and support of antiterrorism policies. *American Journal of Political Science, 49*(3), 593–608.

Huddy, L., Khatib, N., & Capelos, T. (2002). Trends: Reactions to the terrorist atacks of September 11, 2001. *Public Opinion Quarterly, 66*(3), 418–450.

Jacobson, L. (2016, November 16). Is ISIS an 'existential threat' to the United States? *Politifact.* Retrieved from http://www.politifact.com/truth-o-meter/article/2015/nov/16/isis-existential-threat-united-states/

Kagan, F. W., Kagan, K., Cafarella, J., Gambhir, H., & Zimmerman, K. (2016). *U.S. grand strategy: Destroying ISIS and Al Qaeda, report one. Institute for the study of war.* Retrieved from http://www.understandingwar.org/backgrounder/new-report-series-us-grand-strategy-destroying-isis-and-al-qaeda-0

Lee, T. B. (2013, June 12). Here's everything we know about PRISM to date. *The Washington Post.* Retrieved from http://www.washingtonpost.com/

Lewis, S. C., & Reese, S. D. (2009). What is the war on terror? Framing through the eyes of journalist. *Journalism & Mass Communication Quarterly, 86*(1), 85–102.

Lischka, J. A. (2017). Explicit terror prevention versus vague civil liberty: How the UK broadcasting news (de)legitimatise online mass surveillance since Edward Snowden's revelations. *Information, Communication & Society, 20*(5), 665–682.

Lyon, D. (2014). Surveillance, snowden, and big data: Capacities, consequences, critique. *Big Data & Society, 1*(2), 1–13. doi:10.1177/2053951714541861

Lyon, D. (2015). *Surveillance after Snowden.* Malden, MA: Polity Press.

McStay, A. (2017). *Privacy and the media.* Thousand Oaks, CA: Sage.

Mols, A., & Janssen, S. (2017). Not interesting enough to be followed by the NSA. *Digital Journalism, 5*(3), 277–298.

Morin, A. (2016). Framing terror: The strategies newspapers use to frame an act as terror or crime. *Journalism & Mass Communication Quarterly, 93,* 986–1005. doi:10.1177/1077699016660720

Mueller, J. E. (2006). *Overblown: How politicians and terrorism industry inflate national security threats and why we believe them.* New York, NY: Free Press.

Mueller, J. E., & Stewart, M. (2015). *Chasing ghosts: The policing of terrorism.* New York, NY: Oxford University Press.

O'Connor, A. J., & Jahan, F. (2014). American Muslims' emotional and behavioral responses to government surveillance. *Journal of Muslim Mental Health, 8*(1).

Peffley, M., Knigge, P., & Hurwitz, J. (2001). A multiple values model of political tolerance. *Political Research Quarterly, 54*(2), 379–406.

Penney, J. W. (2016). Chilling effects: Online surveillance and wikipedia use. *Berkeley Technology Law Journal, 31*(1), 117–172.

Pew Research (2015, May 29). *What Americans think about NSA surveillance, national security and privacy.* Retrieved from: http://www.pewresearch.org/fact-tank/2015/05/29/what-americans-think-about-nsa-surveillance-national-security-and-privacy/

Pyszczynski, T., Abdollahi, A., Solomon, S., Greenberg, J., Cohen, F., & Weise, D. (2006). Mortality salience, martyrdom, and military might: The great satan versus the axis of evil. *Personality and Social Psychology Bulletin, 32*, 525–537.

Rainie, L., & Madden, M. (2015, March 16). *Americans' privacy strategies post-Snowden.* Pew Research Center, Internet Science and Tech. Retrieved from http://www.pewinternet. org/2015/03/ 16/americans-privacy-strategies-post-snowden/

Rainie, L., & Maniam, S. (2016, February 19). *Americans feel the tensions between privacy and security concerns.* Retrieved from http://www.pewresearch.org/fact-tank/2016/02/19/americans-feel-the-tensions-between-privacy-and-security-concerns/

Reddick, C. G., Chatfield, A. T., & Jaramillo, P. A. (2015). Public opinion on National Security Agency surveillance programs: A multi-method approach. *Government Information Quarterly, 32* (2), 129–141.

Reynolds, A., & Barnett, B. (2003). This just in ... How national TV news handled the breaking "live" coverage of September 11. *Journalism & Mass Communication Quarterly, 80*(3), 689–703.

Roskos-Ewoldsen, D. R., Roskos-Ewoldsen, B., & Dillman Carpentier, F. (2009). Media prim- ing: An updated synthesis. In J. Bryant & M. B. Oliver (Eds.), *Media effects: Advances in theory and research* (pp. 74–93). New York, NY: Routledge.

Ross, S. D., & Bantimaroudis, P. (2006). Frame shifts and catastrophic events: The attacks of September 11, 2001, and *New York Times*'s portrayals of Arafat and Sharon. *Mass Communication & Society, 9*(1), 85–101.

Rucker, D. D., Preacher, K. J., Tormala, Z. L., & Petty, R. E. (2011). Mediation analysis in social psychology: Current practices and new recommendations. *Social and Personality Psychology Compass, 5/6*, 359–371.

Saleem, M., Prot, S., Anderson, C. A., & Lemieux, A. F. (2015). Exposure to Muslims in media and support for public policies harming Muslims. *Communication Research.*

Scheufele, D. A., Nisbet, M. C., & Ostman, R. E. (2005). September 11 news coverage, public opinion, and support for civil liberties. *Mass Communication and Society, 8*(3), 197–218.

Stouffer, S. A. (1955). *Communism, conformity, and civil liberties: A cross-section of the Nation speaks its mind.* Garden City, NY: Doubleday.

Stoycheff, E. (2016). Under surveillance: Examining facebook's spiral of silence effects in the wake of NSA internet monitoring. *Journalism & Mass Communication Quarterly, 93*(2), 296–311.

Sullivan, J. L., Piereson, J., & Marcus, G. E. (1982). *Political tolerance and American democracy.* Chicago, IL: University of Chicago Press.

Vultee, F. (2010). Securitization: A new approach to the framing of the "war on terror." *Journalism Practice, 4*(1), 33–47.

Vultee, F., Lukacovic, M., & Stouffer, R. (2015). Eyes 1, brain 0: Securitization in text, image and news topic. *International Communication Research Journal, 50*(2), 111–138.

White, C., Duck, J. M., & Newcombe, P. A. (2012). The impact of media reliance on the role of perceived threat in predicting tolerance of Muslim cultural practice. *Journal of Applied Social Psychology, 42*(12), 3051–3082.

Yarchi, M., Wolfsfeld, G., Sheafer, T., & Shenhav, S. R. (2013). Promoting stories about terrorism to the international news media: A study of public diplomacy. *Media, War & Conflict, 6*(3), 263–278.

APPENDIX

Participants in the perceived surveillance condition were presented with the following message on the bottom of the introductory screen of the study experiment.

Activity on this site may be monitored by the U.S. National Security Agency

The Impact of Terrorist Attack News on Moral Intuitions and Outgroup Prejudice

Ron Tamborini

Matthias Hofer

Ron Tamborini (Ph.D., Indiana University, 1982) is a Professor in the Department of Communication at Michigan State University. His research interests include how characteristics of traditional and new media alter psychological experience and influence users.

Matthias Hofer (Ph.D., University of Zurich, 2013) is a senior research and teaching associate at the Institute of Mass Communication and Media Research, University of Zurich. His research interests include media audiences and effects, entertainment and emotion research, presence research, and media effects through the lifespan.

Sujay Prabhu (M.A., University of Mumbai, 2007) is a doctoral candidate at Michigan State University. His research interests include the role of intuitive motivations in the influence and evaluation of media as well as the development of implicit measures that can accurately gauge the accessibility of preconscious concepts.

Clare Grall (M.A., Michigan State University, 2015) is a doctoral student in the Department of Communication at Michigan State University. Her research interests include the cognitive processing of and affective responses to narrative entertainment.

Eric Robert Novotny (M.A., University of Buffalo, 2014) is a doctoral student in the Department of Communication at Michigan State University. His research interests include the use of virtual environments in facilitating interpersonal synchrony between humans and virtual agents and the real–world behavioral outcomes of this phenomenon.

Lindsay Hahn (M.A., Kent State University, 2013) is a doctoral candidate in the Department of Communication at Michigan State University. Her research interests include the cognitive processes surrounding narrative media experience and influence in both children and adults.

Brian Klebig (M.A., University of Central Florida, 2014) is a doctoral student in the Department of Communication at Michigan State University. His research interests include pre-conscious drives that influence subsequent thought and behavior.

Sujay Prabhu, Clare Grall, Eric Robert Novotny, Lindsay Hahn, and Brian Klebig

Using logic suggested by the model of intuitive morality and exemplars, we examined the impact of exposure to terrorist attack news coverage on the salience of moral intuitions and prosocial behavioral intentions toward outgroup members. In an experiment, participants were randomly assigned to watch news of the 2015 Paris terrorist attacks or a control news story. Afterward, we measured the salience of five moral intuitions (sensitivity to care, fairness, loyalty, authority, and purity) and the participants' prejudice (i.e., the lack of intentions to help outgroup members). Results showed that exposure to terrorist attack news (a) increased the salience of respect for authority and subsequently (b) reduced prosocial behavioral intentions toward outgroup members. Closer inspection revealed that authority salience mediated the effect of terrorist news exposure on these behavioral intentions toward outgroup members. In a second study using the same design as in the first study, we ensured that the ingroup and the outgroup addressed in the first study were indeed perceived differently on dimensions of ingroup membership.

On the evening of November 13, 2015, a series of coordinated terrorist attacks took place in Paris. The attacks began at 9:20 p.m. local time and consisted of several suicide bombings and mass shootings. At least 126 civilians were killed and 389 were injured. A terrorist group known as Islamic State (commonly referred to as ISIS) claimed responsibility for these attacks. ISIS declared the attacks as a response to French airstrikes against ISIS's militants in Syria. This marked the deadliest attack on the nation of France since World War II (Alderman & Ardley, 2015).

Terrorist attacks like those in Paris not only are a national tragedy but also create an international crisis when they are perceived to threaten the core values that underlie society (Mogensen, 2008). Accordingly, such attacks typically receive extensive media coverage worldwide and increase the public's media use during and after such crises (Althaus, 2002). Media coverage associated with terrorist attacks can dramatically change public opinion and attitudes, as such attacks "manifest themselves in the minds of people as a threat to personal and national security" (Boomgaarden & De Vreese, 2007, p. 355; see also Traugott et al., 2002).

In particular, exposure to terrorist news has been shown to worsen attitudes toward outgroup members who are perceived to be associated with the attacks. For example, research has shown that exposure to news about Dutch filmmaker Theo Van Gogh's murder by Islamic terrorists ultimately increased antisocial attitudes toward Muslims (Das, Bushman, Bezemer, Kerkhof, & Vermeulen, 2009). Similar research showed higher anti-Arabic prejudices after the 2004 Madrid train bombings (Echebarria-Echabe & Fernández-Guede, 2006). A spike in anti-Islamic hate crimes in the United States after 9/11 suggests that such prejudice is not restricted to attitudes but extends to behavior as well (Byers & Jones, 2007).

Although prior research has been useful in identifying the effects of terrorist attacks on explicit attitudes such as outgroup prejudice, little is known about the cognitive processes that can facilitate these outcomes (but see Das et al., 2009). Concern with media's ability to influence social attitudes, stereotypes, and prejudicial behavior have dominated the attention of media researchers and social critics alike, but without an understanding of the cognitive mechanisms that govern these outcomes, we have little chance of solving their related social problems. We argue that the influence of terrorist news exposure on explicit attitudes and prejudices is mediated by the accessibility of moral intuitions (i.e., gut instinct mechanisms related to judgments of morality). More specifically, in the present study we apply the mediating processes outlined in the model of intuitive morality and exemplars (MIME; Tamborini, 2011, 2013) to argue that news coverage of the 2015 Paris terrorist attacks can affect the salience of evolutionarily developed group-centric moral intuitions that, in turn, decrease prosocial behavioral intentions toward outgroup members. We interpret this decrease in prosocial intentions towards outgroup members as an indicator of outgroup bias. In addition, we explore how these effects might be a function of the concreteness of news stories.

THE MIME

The MIME offers a framework for extending traditional approaches to media effects research to increase understanding of the manner in which terrorist attack news can impact social perceptions and outgroup prejudice. Models that outline the relationship between media exposure and social behavior are abundant; however, the MIME is distinguished by its focus on the role of *moral intuitions*. More specifically, the MIME suggests that media content can influence evolutionarily developed moral instincts (known as moral intuitions) in audience members and outlines the short-term and long-term processes through which moral judgment occurs as a function of these instincts/intuitions. According to the MIME, in the short term, exposure to morally relevant environmental cues (including media exemplars) can increase the temporary salience (i.e., accessibility) of related intuitions. Once activated, these moral intuitions can impact attitudes, decision making, and subsequent behavior. The MIME proposes that in the long term, repeated exposure to morally relevant media content can make the respective intuitions chronically salient and, through this mechanism, affect enduring behavior. The MIME combines exemplification theory (Zillmann, 2002) and moral foundations theory (MFT; Haidt & Joseph, 2007) to make specific predictions about how media exemplars activate individual moral intuitions.

Moral Intuitions

The MIME draws moral intuitions from MFT, which describes these instincts as communally beneficial "bits of mental structure" (Haidt & Joseph, 2007, p. 381) that cause instinctive, gut reactions of right/wrong in response to specific actions or behaviors. MFT contends that these intuitions have developed over thousands of years through the course of evolution and have aided the survival of human beings. Unlike accounts that define moral judgment as a product of rational cognitive processes (e.g., Kohlberg, Levine, & Hewer, 1983), MFT regards moral judgment as the result of intuitive affect. Proponents of the theory argue that judgments are most often driven by moral intuition, which form the foundation for and automatically guide gut determinations of right or wrong without the need for deliberative thinking (Haidt & Joseph, 2007).

MFT identifies five of these intuitions, each associated with a sensitivity to a specific domain of social behavior: (a) *care* is based on sensitivity to the suffering of others and the resulting empathic responses; (b) *fairness* pertains to principles related to reciprocity, such as equal treatment or equitable distribution of resources; (c) *ingroup loyalty* deals with commitment and favoritism toward ingroup members; (d) *authority* is related to respect for and deference to traditions and hierarchies; and (e) *purity* is concerned with noble living and disgust mechanisms.

Research by Graham et al. (2011) shows that these five moral intuitions can be combined into two higher order concepts. That is, care and fairness represent so-called *individualizing* intuitions (associated with concern for the well-being of other individuals), whereas loyalty, authority, and purity represent so-called *binding* intuitions (associated with concern for community well-being). The salience of individualizing intuitions has been associated with liberal values, whereas the salience of binding intuitions has been associated with conservative values and attitudes (Graham, Haidt, & Nosek, 2009; Van Leeuwen & Park, 2009). We argue that these intuitions act as mediators through which terrorist news exposure impacts prejudice toward outgroup members.

EFFECT OF NEWS ON TERRORIST ATTACKS ON OUTGROUP PREJUDICE

Prior investigations examining the influence of terrorist attack news have focused on how coverage can affect prejudice toward outgroup members (Das et al., 2009;

Echebarria-Echabe & Fernández-Guede, 2006). However, this research has either not examined any mediating variables or focused on a narrow range of mediating mechanisms responsible for such change. For instance, Das et al. (2009) proposed the mediating mechanisms described in terror management theory (TMT; Greenberg, Pyszczynski, & Solomon, 1986) to explain the influence of terrorist news on outgroup prejudice. More specifically, this research reasoned that terrorist news exposure increased death-related thoughts, which made people sublimate mortality by defending their enduring cultural worldviews. In doing so, TMT explains, people become suspicious of any who pose a threat to these worldviews.

However, there are limits to TMT. For instance, although TMT is a useful framework for investigating the effects of terrorist news on outgroup prejudice (Das et al., 2009), the cognitive mechanisms that mediate the effect of terrorist news on these outcomes have been overlooked in most research or difficult to interpret when they have been examined (e.g., Das et al., 2009). Other studies generally claim that fear or anxiety about death (hence the term *anxiety buffer*) is the mechanism mediating the outcomes (see Burke et al., 2010, pp. 159–176); however, these studies refrain from measuring fear of death. This intentional omission makes sense. The scale itself would be a reminder of death and, as such, would make a true control condition impossible.

In the context of terrorist news, TMT would explicate how terrorist news sparks death-related thoughts in media users, which in turn lead to defense of cultural worldviews (manifested as outgroup prejudice in this case). Although it is plausible that death-related thoughts can lead to outgroup prejudice, we do not believe that this is the only factor that can contribute to outgroup prejudice. In this article we outline a completely different psychological mechanism (independent of the cognitive mechanism outlined in TMT) through which terrorist news can affect prejudice toward outgroups. We have attempted to explain how these instincts can lead to the spontaneous activation of outgroup prejudice, which do not involve personal fear of death per se.

Whereas TMT research provides evidence of outcomes stemming from an individual's fear of his or her own death (e.g., Florian & Mikulincer, 1997), it is less applicable to outcomes driven by threat to others. One could argue that terrorist news, especially when involving a terrorist attack in another country, evokes thoughts of threats to other people and the social order more than thoughts of one's own death, as mortality salience evokes. Therefore, a theory that explains the influence of terrorist news on mechanisms associated with societal outcomes would be more applicable. Instead of TMT we use the MIME, which combines exemplification theory and MFT to predict the effect of terrorist news on the accessibility of binding moral intuitions and subsequent outgroup bias.

Although recognizing that fear of death or other intuitions related to the self may play a role in determining news of terrorist attacks effects outgroup prejudice, we contend that the group-related binding intuitions, which MFT describes as being present in all human beings from birth, are (a) sensitive to media stimuli, which can provoke a sense of threat to the group, and (b) capable of influencing subsequent attitudinal and behavioral outcomes. Thus, this study is intended to test if the binding intuitions can serve as mediators of the influence of terrorist-related news on outgroup bias. Notably, previous research has been limited to examining how terrorist attack news affects attitudinal responses such as hatred of outgroup members, without addressing its impact on behavioral intentions. The present study extends this research to examine the effects of terrorist news coverage on prosocial behavioral intentions toward outgroup members (indicating benevolent concern) or the lack thereof (indicating malevolent, antisocial prejudice).

HYPOTHESIZED MODEL

The study's central hypothesis predicts that the salience of the binding intuitions will mediate the influence of exposure to terrorist attack news on prosocial behavioral intentions toward outgroups. More specifically, we hypothesize the following:

H1a: Exposure to news of the Paris terrorist attacks will increase the salience of binding intuitions of loyalty, authority, and purity.

Support for this logic can be found in work by Van Leeuwen and Park (2009), who showed that the impact of perceptions of social danger on conservative attitudes was mediated by the salience of binding intuitions. In showing this mediation, they established a causal link between fear and the activation of the binding intuitions. They explained that binding intuitions offered "protection" (p. 170) against threat, presumably because humans have benefited from group affiliation when their survival is threatened. Given that these moral intuitions are described as evolutionarily developed instincts, it is reasonable to believe that natural selection has favored the development of instincts that bind an individual to his or her group in the face of threats, with one modern manifestation being terrorist attacks.

This evolutionary account is explained more thoroughly by Van Leeuwen, Park, Koenig, and Graham (2012), who showed that pathogen prevalence in countries strongly predicted the salience of the three binding intuitions for people living in these countries. They specifically discuss how pathogen contamination poses a threat to survival, and how human beings with strong group-related instincts sensitive to threat may have been more likely to survive and pass on those instincts to future generations. More specifically, commitment and bonding with fellow ingroup members can offer protection from dangerous outside

71

threats, so it is quite likely that natural selection favored humans whose ingroup loyalty intuition was immediately activated upon perceived threat. In addition, adherence to traditions and taking orders (authority) in response to threat, and disgust mechanisms that can make people wary of foreign and potentially harmful objects (purity) in the face of threat have evolutionary advantages that likely made the authority and purity intuitions sensitive to threat over the course of human history.

In the context of modern times, it is arguable that few events can stimulate the primordial sense of threat to survival as a terrorist attack can (Boomgaarden & De Vreese, 2007). Hence, we could expect that exposure to mediated threat in the form of terrorist attack news should innately activate the three binding intuitions in human beings. Previous research on the effects of TV news exposure on binding intuitions supports this prediction. For instance, exposure to news content focusing on nuclear threat was found to increase the salience of authority without influencing other intuitions (Tamborini, Prabhu, Hahn, Idzik, & Wang, 2014; Tamborini, Prabhu, Wang, & Grizzard, 2013).

Although MIME research has not yet explored the role of binding intuitions on behavioral outcomes, there is reason to believe that all three binding intuitions can influence responses to outgroup members. The mechanisms that underlie the binding intuitions lead humans not only to putting communal needs above their own but also to increasing bias toward non-community members in times of threat (Kugler, Jost, & Noorbaloochi, 2014). Extending Haidt and Joseph (2007), we argue that activation of the ingroup loyalty intuition can lead to higher levels of commitment to one's ingroup, which might inhibit active participation with outgroups. The authority intuition, once primed, increases the salience of rules that protect the group including preferential treatment of ingroup over outgroup members, and submission to authoritative rules that may protect from outgroup members. Finally, the increased accessibility of the purity intuition can lead to disgust and avoidance responses directed toward outgroup members who might represent physical and spiritual contamination.

In line with the logic of the MIME, we predict the following:

H1b: The salience of binding intuitions will decrease prosocial behavioral intentions toward outgroup members (which we consider an *increase* in prejudice toward outgroup members).

Finally, our third hypothesis (H1c) represents the mediation (by binding intuition salience) of the effect of exposure to terrorist attack news on prosocial behavioral intentions toward outgroup members.

H1c: Binding intuitions will mediate the effect of exposure to terrorist attack news on prosocial intentions toward outgroup members.

The MIME predicts that content features in media will prime intuition salience to influence decisions and related behaviors (Tamborini, 2013). Although previous research has separately examined paths H1a and H1b, to our knowledge no research to date has examined the mediation process (H1c) outlined in the MIME. All three hypotheses are represented in the research model in Figure 1.[1]

THE IMPORTANCE OF EXEMPLAR STRENGTH OF TERRORIST NEWS STORIES

In addition to MFT, the MIME also draws on logic from exemplification theory (Zillmann, 2002) to explain how specific media content can *exemplify* moral principles, and hence serve as *exemplars* of moral intuitions. An exemplar is a prototypical example of some phenomenon that demonstrates the important aspects of that phenomenon. For instance, a media exemplar of the care intuition would be a video clip of the moment a mother feeds her hungry child

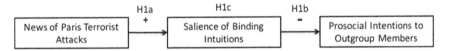

FIGURE 1 A mediation model that highlights the role of binding intuition salience as a mechanism through which exposure to terrorist attack news influences prosocial behavioral intentions toward outgroup members.

(Tamborini, Prabhu, Lewis, Grizzard, & Eden, 2016). Such an exemplar would

---

[1] It is possible that the salience of individualizing intuitions (care and fairness) could mediate the influence of exposure to media exemplars on behavior in much the same way as predicted in our model with binding intuitions. However, inconsistencies in previous research relating exposure to terrorist attack news to individualizing intuitions make expectations for this model difficult to ascertain. One could argue that concern for the victims of the attacks would make the care intuition more salient. Grizzard et al.'s (2015) findings showed that, in addition to influencing the salience of the binding intuitions, exposure to terrorist film clips also increased the salience of the more liberal, or individualizing, intuitions. Conversely, other research has shown that although knowledge of terrorist attacks increases conservative and authoritarian attitudes related to the binding intuitions, it can decrease liberal attitudes related to the individualizing intuitions. For example, Bozzoli and Müller (2011) found that the 2005 London bombings significantly decreased support for civil liberties while increasing support for enhanced security. Given the contradictory findings, there are no solid grounds for a hypothesis regarding the effects of exposure to terrorist activities on individualizing moral intuitions. As such, we can make no predictions about the ability of individualizing intuitions to mediate the influence of exposure to terrorist attack news on intention to donate to outgroup members. Nevertheless, although we pose no specific hypothesis here, as described next, individualizing intuitions are measured and examined in our study.

temporarily increase the accessibility of the care intuition in viewers. Although considerable research demonstrates the importance of mere exposure to media exemplars as a determinant of media effects (e.g., Zillmann & Brosius, 2000), less research has examined the role of the *strength* of a media exemplar, which relates to how *concrete* and *emotion inducing* the exemplar is.

In the context of terrorism-related media, in several previous studies on terrorist news, outgroup prejudice has been attributed to simply being made aware of a potential attack (e.g., Boomgaarden & De Vreese, 2007; Bozzoli & Müller, 2011; Echebarria-Echabe & Fernández-Guede, 2006). We know of only one study that examines characteristics of exemplar strength in this context, in particular, the level of graphicness portrayed (Grizzard et al., 2015). We apply components of the MIME to more closely examine the possible effects of exemplar strength in terrorist news on intuition salience and subsequent outgroup prejudice.

As just mentioned, the MIME adopts exemplification logic to argue that a media exemplar's impact on judgment increases as the exemplar becomes stronger (i.e., more concrete and emotional; Zillmann, 2002). Support for this argument is widespread in research examining the effect of exemplification in news (Zillmann & Brosius, 2000). Evidence for the effects of concrete exemplars can be found in other literatures. For example, McKinley and Fahmy (2011) showed that the level of graphicness (a characteristic tied to concreteness and emotionality) of images of the Israeli–Palestinian conflict affects viewers' emotional states. Grizzard et al. (2015) specifically examined the influence of graphicness in news footage of terrorist acts (i.e., an ISIS execution) on moral emotions and subsequent intuition salience. They found that the graphicness of the depiction of a mass shooting affected the experience of anger and disgust, which then increased the salience of moral intuitions. This leads to our second hypothesis (H2):

H2: The positive influence of terrorist-news stories on the salience of binding intuitions will be stronger when they are more concrete/emotional.

STUDY 1

Method

Participants. There were 238 undergraduate students[2] at a large university in the midwestern United States who participated in a between-subjects experiment ($n_{female} = 158$, $M_{age} = 20.04$, $SD = 1.67$). All students in the sample received course

---

[2] One important reason for using a sample of undergraduate communication students was that student samples are quite homogenous in terms of age and developmental stage. This homogeneity is important when it comes to basic research because it ensures internal validity of results (see, e.g., Greenberg, 1987).

credit for their participation. Participants were randomly assigned to one of three experimental conditions ($n_{control} = 77$, $n_{low\ exemplar\ strength} = 77$, $n_{high\ exemplar\ strength} = 84$). The university's ethics review board approved all procedures.[3]

Procedure. The study began 2 weeks after the 2015 Paris terrorist attacks and was completed within 1 week. To complete the study, participants sat in front of a 17-in. computer screen. After signing an online consent form, participants were randomly assigned to one of six experimental conditions in a 3×2 design that varied exposure to news of the Paris terrorist attacks (high exemplar strength, low exemplar strength, control) and target of prosocial behavioral intentions (ingroup vs. outgroup). Next they completed measures assessing the salience of moral intuitions and prosocial behavior toward the target group. Participants were then debriefed and dismissed. The target of prosocial behavior was manipulated in order to determine whether the influence of intuition salience on prosocial behavioral intentions (H1b) varied as a function of the target of the prosocial behavioral intentions. More precisely, we expected intuition salience to affect prosocial behavioral intentions only for outgroup members.

Stimuli. News stories were drawn from various online sources. The TV reports were disguised as coming from the same network (CBS News). Participants in the control group saw video news stories about a restaurant and about a new method of electricity production using the kinetic energy of sea waves. In addition to the control group, we created two treatment conditions varying exemplar strength. Participants in the treatment groups saw the same two news stories seen by the control group, and a third story that was either (a) video footage of the Paris attacks with a voice-over (high exemplar strength) or (b) a video consisting of still images selected from the same video with the same voice-over (low exemplar strength). Both treatment conditions showing the Paris attacks depicted a time line of events superimposed over a map of locations in the city where attacks took place, and interspersed recordings from surveillance cameras and private cell phones. For example, in the high exemplar condition, participants heard the newscaster describe a scene in which shattering glass flies across the screen, a gunman is seen approaching café patrons, chairs are being tossed, and people are seen diving for cover under tables, while seeing moving

---

[3] This study was deemed exempt under Institutional Review Board (IRB) ID# i04990 on October 29, 2015, because (a) human subjects could not be identified, directly or through identifiers linked to the subjects, and (b) disclosure of human subjects' responses outside the research could not reasonably place the subjects at risk of criminal or civil liability or be damaging to the subjects' financial standing, employability, or reputation. Approval occurred before the terrorist attacks. We contacted IRB to approve changes to our stimuli in an existing exempt IRB application (procedures remained the same). The IRB informed us that the changes did not alter the exempt status.

images of these events. In the low exemplar condition, participants heard the same description while seeing still shots of shattered glass, the gunman, overturned chairs, and people crouched under tables.

The use of videos versus still images is expected to vary exemplar strength both in terms of perceived concreteness and emotionality. Concreteness is expected to be higher in the video condition because participants could see all the character and object movements that the newscaster was describing verbally. Emotionality is expected to be higher in the video condition because of its structural features. Previous research on structural features shows that motion in media stimuli increases both the self-reported emotionality of the media content and physiological arousal in respondents (Simons, Detenber, Roedema, & Reiss, 1999). For example, things like zooms, pans, and other forms of camera action, as well as action within a shot, can increase arousal. This increase in arousal, in association with the content of terrorist attacks, is expected to heighten the perceived emotionality of the video.

Measures

Moral Intuition Salience. The salience of five moral intuitions was measured by items the moral foundations questionnaire (MFQ; Graham et al., 2011). The MFQ consists of 30 questions with six items for each moral intuition. More precisely, each moral intuition is assessed by three relevance statements (each statement beginning with "When you decide whether something is right or wrong, to what extent are the following considerations relevant to your thinking?") and three judgment statements (each statement beginning with "Please read the following sentences and indicate your agreement or disagreement."). Participants indicated their agreement and disagreement on a 7-point Likert scale from 1 (*not at all relevant*) to 7 (*extremely relevant*) the relevance scales and 1 (*strongly disagree*) to 7 (*strongly agree*) for the judgment scales.

Prosocial Behavioral Intentions. The intention to donate measure contained a target manipulation (ingroup vs. outgroup). That is, participants were first asked whether they would be willing to increase their tuition fees the following semester to support a program seeking to eliminate childhood hunger either in the Middle East (target: outgroup, $n = 120$) or in the United States (target: ingroup, $n = 118$). If they did not agree, they were coded as $0. If they agreed, they were asked whether they would be willing to have between $1 and $10 added to their tuition the following semester. The range of $1–$10 was selected based on consideration of the typically limited finances of college students. The means and standard deviations of both the MFQ and the donation amounts are presented in Table 1, along with their zero-order correlations.

TABLE 1
Means, Standard Deviations, and Zero-Order Correlations

	α	M	SD	1	2	3	4	5
1. Care	.63	4.43	0.70					
2. Fairness	.65	4.29	0.65	.62**				
3. Loyalty	.61	3.84	0.72	.25**	.28**			
4. Authority	.56	4.00	0.67	.28**	.28**	.62**		
5. Purity	.71	3.66	0.87	.31**	.26**	.52**	.60**	
6. Amount donated	—	3.65	4.20	.13*	.14*	−.09	−.11	−.08

Note. * $p < .05$. **$p < .01$.

Manipulation Check. Finally, to examine whether the exemplar strength manipulation was successful, we asked participants how concrete the events in the videos seemed to them ("How clear and concrete did the events in the video seem to you?"). The concreteness measure used a 7-point scale offering itemized responses that varied along a continuum from 1 (*so unclear it was incomprehensible*) to 7 (*so clear it was like you were there; M = 5.51, SD = 0.93*). We also asked participants how emotional they perceived the events in the videos ("How emotional did the video scenes seem to you?"), on a scale from 1 (*very unemotional*) to 7 (*very emotional; M = 4.65, SD = 1.68*).

Results

To examine the success of our exemplar strength induction, we conducted two separate one-way analyses of variance (ANOVAs) on items asking participants to evaluate the emotionality and concreteness of the terrorist news videos. The ANOVA examining emotionality showed a significant overall effect of the induction on perceived emotionality, $F(2, 235) = 72.00$, $p < .001$, $\eta_p^2 = .38$. A post hoc Scheffé test revealed that participants in the control condition showed significantly lower values on the emotionality measure ($M = 3.16$, $SD = 1.33$) than both participants in the low ($M = 5.35$, $SD = 1.30$) and high ($M = 5.38$, $SD = 1.36$, $p < .001$) exemplar strength conditions. However, the low and high conditions did not differ significantly. In terms of perceived concreteness, we found a significant difference between the low ($M = 5.21$, $SD = 1.15$) and high ($M = 5.62$, $SD = 0.90$, $p = .01$) concreteness conditions.

To examine our hypotheses we first conducted a multivariate analysis of variance with the experimental factor (control vs. low vs. high exemplar strength) as the independent variable and the salience of the five moral intuitions as the dependent variables. This analysis revealed that the multivariate effect of the experimental manipulation on moral intuition salience approached significance: Wilks's $\Lambda = .93$ $F(10, 462) = 1.73$, $p = .07$, $\eta_p^2 = .04$.

Post hoc analyses using the Scheffé post hoc criterion for significance indicated that there was no difference between the two exemplar strength conditions on moral intuitions. H2 was accordingly not supported. Because of this, and our observation that the exemplar strength conditions did not differ on emotionality, we collapsed the high and low exemplar strength conditions into a single experimental condition, called terrorist news exposure, for all subsequent analyses.

Follow-up ANOVAs showed that only authority and purity salience differed between the experimental conditions. Specifically, we found that terrorist news exposure had a significant effect on the salience of authority, $F(2, 235) = 4.57$, $p < .05$, $\eta_p^2 = .04$, and that the effect of terrorist news exposure on the salience of purity approached significance, $F(2, 235) = 2.89$, $p = .058$, $\eta_p^2 = .02$. We found no effect of the terrorist news video on loyalty or the two individualizing intuitions, care and fairness (all $F < 1$).

To test H1a we conducted analyses of variance examining the effect of terrorist news exposure on binding intuition salience (for authority, loyalty, and purity). The salience of authority was significantly lower in the control condition ($M = 3.82$, $SD = .75$) than in the terrorist news condition ($M = 4.10$, $SD = .61$), $F(1, 236) = 8.84$, $p = .01$, $\eta_p^2 = .04$. The same pattern was found for purity, as the salience of this moral intuition was lower in the control condition ($M = 3.46$, $SD = .84$) than in the terrorist news condition ($M = 3.75$, $SD = .88$), $F(1, 236) = 5.67$, $p = .02$, $\eta_p^2 = .02$. No significant effect was found for the salience of loyalty, $F < 1$. Therefore, H1a was partially supported by our data, as significant results were found for the salience of authority and purity, but not for loyalty.

To test H1b and H1c, we estimated three moderated mediation models to examine the effects of the two moral intuitions that were affected by terrorist news exposure (authority and purity). For all three models, donation intention target (ingroup vs. outgroup) served as the moderator in order to determine whether the influence of intuition salience on prosocial behavioral intentions varied as a function of target-group membership. In the first model we tested for the mediational effect of purity and authority salience combined, and in the second and the third models we tested for the mediational effect of purity salience and authority salience separately.

To estimate these models, we used the PROCESS macro by Hayes (2013; Model 15). Only the model with authority salience as the mediator yielded significant results (see Table 2). As can be seen from Table 2, terrorist news exposure had an effect on the authority intuition, such that participants who saw footage of the Paris attacks showed higher levels of authority salience than participants who saw the control videos ($b = 0.27$, $SE = .09$, $p = .01$). Authority intuition salience, in turn, negatively affected the amount participants were willing to donate ($b = -1.27$, $SE = .60$, $p = .03$) and so did terrorist news exposure ($b = -2.21$, $SE = .84$, $p = .01$). However, the effect of authority

TABLE 2

Moderated Mediation Model: Indirect Effect of Terrorist News (IV) on Donation Amount Offered (DV), Mediated by Authority Salience (M) Moderated by Ingroup versus Outgroup (Mo)

Predictors	Mediator Variable Model (DV = Authority)			
	b	SE	t	p
Terrorist News (0 = control news)	0.27	0.09	2.97	.00
Model summary $R^2 = .04$, $F(1, 236) = 8.83$, $p = .00$				
Dependent Variable Model (DV = Intended Donation Amount)				
Terrorist news (0 = control news)	-2.21	0.84	-2.63	.01
Authority	-1.27	0.60	-2.12	.03
Ingroup vs. outgroup (0 = Middle East)	-7.39	3.27	-2.26	.02
Authority × Ingroup vs. Outgroup	1.42	0.82	1.74	.08
Terrorist News × Ingroup vs. Outgroup	1.76	1.17	1.51	.13
Model summary $R^2 = .06$, $F(5, 232) = 2.99$, $p = .01$				

Conditional Direct Effects of Terrorist News on Intended Donation Amount at Values of Ingroup vs. Outgroup

Ingroup vs. Outgroup	b	SE	t	p
Middle East	-2.21	0.84	-2.63	.01
USA	-0.45	0.81	-0.56	.57

Conditional Indirect Effects of Terrorist News on Donation at Values of Ingroup vs. Outgroup

Mediator	b	Boot SE	Boot 95% CI
Authority Middle East	-0.34	0.20	[-0.85, -0.04]
Authority USA	0.04	0.17	[-0.25, 0.43]

Note. b = unstandardized coefficient, bootstrap samples = 5,000.

salience on intended donation amount was moderated by ingroup versus out-group ($b = -1.42$, $SE = .82$, $p = .08$). That is, authority salience only affected donation intentions toward outgroup but not ingroup members (see Figure 2), which resulted in a moderated indirect effect. In other words, terrorist news exposure negatively affected prosocial behavioral intentions through authority salience but, as expected, only for the outgroup (i.e., a Middle Eastern organization against child hunger, $b = -0.34$, Boot $SE = 0.20$), Boot CI [−0.85, −0.04]. In terms of a U.S. organization, the indirect effect through authority salience was not significant ($b = 0.04$, Boot $SE = 0.17$), Boot CI [−0.25, 0.43].

Discussion

The findings of the current study suggest that exposure to news of terrorist attacks has the capacity to activate authoritarian intuitions to a level that can lead to outgroup prejudice (i.e., a decrease in prosocial behavioral intentions toward an outgroup organization). They demonstrate that a single exposure to terror attack news can increase the temporary salience of the authority intuition and shape subsequent response toward outgroup members. The results show that exposure to news about the 2015 Paris terrorist attacks significantly increased the salience of some intuitions and not others. More precisely, exposure to news

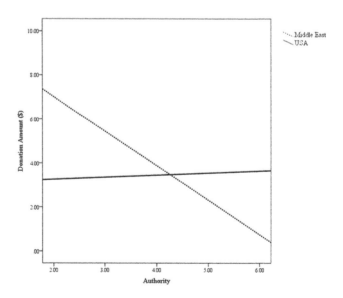

FIGURE 2 Interaction of authority intuition salience and donation target on intended donation amount.

videos of these terrorist attacks increased the salience of two binding moral intuitions, namely, authority and purity. The third binding intuition (loyalty) was not affected by exposure to these videos. This may have been because loyalty as a concept pertains more to small groups (e.g., "tribes, gangs, and teams"; Haidt & Joseph, 2007, p. 17) rather than broad societies as a whole. Moreover, neither of the individualizing intuitions (care and fairness) was affected by the terrorist attack videos. In terms of the exemplar strength of the terrorist attack videos, our manipulation did not vary emotionality as expected and consequently, we did not find any effect of the exemplar strength induction (i.e., concreteness/emotionality) on moral intuition salience. However, perceived concreteness was affected by our exemplar strength induction.

Also, analyses revealed that media exposure had a significant effect on outgroup prejudice. We regarded a decrease in the willingness to donate to outgroup members as an indicator of outgroup prejudice. More precisely, exposure to terrorist attack news significantly decreased the amount of money viewers were willing to donate to an outgroup organization (i.e., a Middle Eastern organization against child hunger) if they were willing to donate at all. In line with our hypothesized model, this effect of terrorist news on donation intention was mediated by intuition salience. Notably, although terrorist news exposure increased the salience of both the authority and purity intuitions, tests showed that the mechanism mediating terrorist news exposure's effect on donation was the salience of authority. Purity did not serve as a mediator of this effect. Finally, the mediated effect of terrorist attack news on willingness to donate was not found for ingroup members.

Altogether, our findings show that exposure to terrorist news can ultimately affect viewers' outgroup prejudice. However, one might argue that in our study the terms *outgroup* and *ingroup* were not properly defined. In other words, we did not ensure whether the target donation group in our study—U.S. children and Middle Eastern children—were perceived differently in terms of perceived ingroup membership. Therefore, we conducted an additional study.

STUDY 2

To address concerns that the target donation groups used in Study 1 (i.e., U.S. children and Middle Eastern children) were indeed perceived differently on dimensions of ingroup membership as expected, a second study was conducted.[4] We hypothesized that participants would rate American targets higher than Middle Eastern targets on measures of perceived ingroup membership.

---

[4] Study 2 was conducted to address reviewer concerns regarding the success of our perceived group membership manipulation in our main study. We want to thank the reviewer for identifying this concern.

Method

Participants.   Participants (N = 266, n_{female} = 202, M_{age} = 20.33, SD_{age} = 2.10) were selected from the same population used in Study 1. Again, students in the sample received course credit for their participation, and the university's ethics review board approved all procedures.

Procedure.   The second study duplicated the 3 × 2 design from Study 1 in order to determine the effect of our concreteness and donation target manipulations on a measure of perceived ingroup membership. Participants were randomly assigned to one of the same three experimental videos (control, low concreteness, and high concreteness) and were asked to contribute to one of two target groups (American or Middle Eastern organization against child hunger). Additional measures added specifically for Study 2 assessed the perceived ingroup membership of the target group to which they were assigned. As in Study 1, participants who viewed the low and high concreteness videos were combined into a single attack video condition.

Ingroup Scale.   All participants completed a perceived ingroup membership scale. The scale contained seven items measuring perceived ingroup membership on a Likert-type scale ranging from 1 (*strongly disagree*) to 7 (*strongly agree*). Five of these items were adapted from Leach et al. (2008; e.g., "I feel a bond with [group]"), and the remaining two were created specifically for this study (i.e., "I perceive [group] children as part of my ingroup," "I perceive hungry [group] children as part of my ingroup").

Respondents were asked to evaluate the perceived ingroup nature only for members of the donation target group to which they were assigned. For example, if you were assigned to the American children donation target group, the perceived ingroup membership scale asked only about perceptions of Americans. However, two versions of the seven-item perceived ingroup membership scale were created for both the American and Middle Eastern children target groups in order to determine whether perceived ingroup membership would differ if the target was a child. Participants were randomly assigned to a version asking if they felt ingroup membership toward either (a) children from the assigned target group or (b) members of the assigned target group without specifying age. The ANOVAs on perceptions of ingroup membership scale scores showed that age specification had no significant impact. As such, we combined data from participants who received versions of the scale that differed by age specifica-tion. This left us with a single measure evaluating perceived ingroup membership such that for participants assigned to the American children donation target group, our seven-item scale measured perceived ingroup membership toward Americans, whereas for participants assigned to the Middle Eastern children target group, our scale measured perceived ingroup membership toward Middle Easterners. The seven items were combined to form a single perceived ingroup membership scale (M = 4.63, SD = 1.29, α = .83).

Results

A 2 (control video versus attack video) x 2 (Americans versus Middle Easterners) factorial ANOVA was conducted on perceived ingroup membership scale scores. A significant main effect was found for target group showing that U.S. targets ($M = 5.32$, $SD = 1.07$) were rated higher than Middle Eastern targets ($M = 3.95$, $SD = 1.10$) on the perceived ingroup membership scale, $F(1, 266) = 110.95$, $p < .001$, $\eta_p^2 = .30$. We found no significant effects for video condition or the interaction between video condition and target group ($Fs > 1$).

Discussion

The findings are consistent with our prediction that participants would rate American targets higher on perceived ingroup membership than Middle Eastern targets. Because the mean for Middle Eastern targets was around the midpoint on our scale ($M = 4.00$), which represents *neither agrees nor disagrees*, participants considered Americans to be part of their ingroup, but not Middle Easterners. This is consistent with our contention that the measure of donation toward American versus Middle Eastern children's groups used in Study 1 was indicative of ingroup bias.

GENERAL DISCUSSION

The outcomes of our investigation have several important implications. We focus on three of them in the following sections: First, our observation that the mediation of terrorist news exposure's impact on behavior occurred specifically through the authority intuition directs our attention to the importance of concepts related to authority for understanding outgroup bias. Second, our evidence for the short-term component of the MIME furthers support for the model's predictions that media exposure affects behavioral outcomes by activating moral intuitions. Third, our failure to find that exemplar strength affected the impact of media content on intuition salience or behavior directs our attention to components of the MIME based on exemplification theory.

The Salience of the Authority Intuition and Outgroup Prejudice

We began our investigation by considering the mediating influence of binding intuitions as a whole. Our expectation was that the salience of the binding intuitions would increase, as these intuitions have to do with the welfare of the group over and above the individual. The hypothesized effect of the media coverage of the 2015 Paris terror attacks on binding intuitions is consistent with previous research on the effects of terrorist attacks (Boomgaarden & De Vreese, 2007; Bozzoli & Müller, 2011; Echebarria-Echabe & Fernández-Guede, 2006). Most interesting in this regard is our demonstration that news coverage of these attacks increased the salience of intuitive respect for authority. This effect is in line with prior research showing that exposure to terrorist attacks increased authoritarian attitudes (Bozzoli & Müller, 2011; Hetherington & Suhay, 2011). However, unlike previous research that focuses on the impact of exposure to terrorism on *explicit attitudes* related to authority, the present study demonstrates that such videos can impact the foundational instincts (*intuitions*) upon which such attitudes may be based. In retrospect, however, it may come as little surprise that the salience of the authority intuition in particular proved to be the mechanism through which the decreases in prosocial behavioral intentions toward outgroup members occurred. Previous research has demonstrated that exposure to news of terrorist attacks can influence authoritarian beliefs (Echebarria-Echabe & Fernández-Guede, 2006). Notably, although previous research links other binding intuitions with both knowledge of terrorist news attacks and outgroup bias, we do not find evidence for the mediating role of the other intuitions, whether binding or individualizing.

We did not observe any effect of the Paris terrorist attack news on the salience of ingroup loyalty. This lack of effect may be attributed to the measure we used: The Loyalty subscale of MFQ (Graham et al., 2011). Although ingroup loyalty is conceptualized as both "recognizing, trusting, and cooperating with [ingroup] members" and "being wary and distrustful of members of other groups" (Haidt & Graham, 2007, p. 105) items in the MFQ focus solely on the aspect pertaining to ingroup benefit such as "love our country," "loyalty to family," and being a "team player," while they ignore aspects pertaining to distrust of outgroup members. Indeed, previous research (Echebarria-Echabe & Fernández-Guede, 2006) found that media content primarily influences the aspect of loyalty that pertains to outgroup distrust rather than ingroup loyalty. Although we do believe that both ingroup loyalty and outgroup distrust are different manifestations of the same underlying instinct, each may be more salient in different contexts. For example, listening to the national anthem is more likely to spark ingroup commitment rather than outgroup mistrust. In the current study's context, given that these terrorist acts were committed by a hostile and seemingly maleficent outgroup, the terrorist attack news was more likely to provoke outgroup distrust, which may not have been properly captured in the items of the MFQ scale.

Support for the MIME's Mediation Processes

A central goal of our study was to test the mediation process outlined in the short-term component of the MIME. More precisely, media exemplars can act as primes to increase the salience of moral intuitions, which then affect decision making. Support for this mechanism was provided in our test showing that the salience of the authority intuition mediated the negative influence of terrorist attack news on the willingness to donate to outgroup members. To our knowledge, this is the first time that the MIME's mediation process has been demonstrated empirically. Although previous research has demonstrated the ability of media exposure to increase the salience of moral intuitions (Grizzard et al., 2015; Tamborini et al. 2014), and the ability of intuition salience to predict behavioral outcomes (Tamborini, Bowman et al., 2016), the results of the present investigation test and support the mediation process in its entirety. In doing so, the present findings not only support the logic underlying the MIME but also show the importance of understanding the role of moral intuition salience as a mechanism through which media can influence behavioral outcomes.

We began by expecting that all binding intuitions would mediate the influence of exposure to terrorist news attacks on prejudice. Instead we found that mediation occurred only through the authority intuition. In line with previous research showing the relationships between knowledge of terrorist attacks, authoritarian beliefs, and outgroup bias, the findings of this study suggest that the terrorist attack news created a desire for strong leadership in the face of threat, which led to the change in outgroup prejudice observed here. The implication is that media's impact on these basal instincts has wide-ranging consequences for behavioral outcomes. In this case, the increased salience of the authority intuition may have heightened a desire for authoritarian leadership and an acceptance of social hierarchies. The outcomes are in line with the MIME's predictions that media content does not just increase the salience of specific issues in the minds of audiences but increases the salience of moral intuitions that, in turn, can influence a broad range and thoughts and behaviors.

If researchers can identify intuitions that can shape desired outcomes, message producers can design content better suited for bringing about desired change. For example, if content increases the salience of the authority intuition, which then decreases prosocial behavior toward outgroup members, perhaps content that *decreases* the salience of the authority intuition could increase prosocial behavior toward outgroup-members. There is still need for research identifying the types of message features that will increase or decrease the salience of intuitions related to specific behavioral outcomes. This research is likely to prove challenging; however, the implications of success in this area are considerable.

The Importance of Exemplar Strength

Exemplar strength was varied by a video condition and a still-image condition but had no influence on any intuition. This is inconsistent with previous research that has shown an effect of concrete/graphic exemplars on emotions. For example, McKinley and Fahmy (2011) found effects of graphicness on viewers' emotions—a finding consistent with that of Grizzard and colleagues (2015). Our failure to replicate this effect is also inconsistent with components of the MIME that are based on exemplification theory. One interpretation of this observation is that exemplar strength (i.e., concreteness) does not play a role in an exemplar's ability to increase the salience of an intuition. However, given the findings by both Grizzard et al. (2015) and McKinley and Fahmy (2011) regarding emotional reactions, this interpretation seems rather implausible. Another interpretation is that our stimuli did not vary exemplar strength effectively. Although these are both possibilities, we think a third possibility might account for the failure to find an effect of this factor: We conducted our research only 2 weeks after the Paris attacks, and at this time it did not take a very strong stimulus to strengthen the emotional impact of the news story and increase the salience of related moral intuitions. Reports from participants suggest that the events were not far from their thoughts before exposure to our stimuli, and many reported that they recently read and thought about the events in the news. As such, any stimulus showing news of the Paris terrorist attacks (i.e., whether high or low in con-creteness) may have been powerful enough to fully activate intuitions related to the events. Evidence of this is apparent in our manipulation check for exemplar strength, which showed that although the high and low conditions varied as expected in perceived concreteness, they did not vary in emotionality.

The likelihood that any other news story on the Paris attacks than the ones we used would have activated the authority intuition and altered behavioral intention is supported by the fact that terrorist news exposure *did* affect intuition salience and subsequent behavioral intentions across the combined exemplar strength conditions. Our data suggest that the story was a powerful (i.e., emotional) exemplar at both levels of our exemplar strength manipulation, and as such was able to affect the salience of moral intuitions as measured by the MFQ.

It is worth noting that the materials used to create the Paris attack news stimuli focused on the attack and not on the suffering of victims. It is possible that stimuli focusing more on victimization would have affected the salience of other intuitions such as care or perhaps even fairness. Such a finding would be consistent with previous research by Tamborini et al. (2014) showing the ability of exposure to news stories on tornado victims to increase the salience of care.

Limitations

Several limitations are worth mentioning. In terms of individualizing intuitions, the observation that terrorist news exposure did not affect the salience of care or fairness shows that these intuitions differed from the authority and purity intui-tions. Perhaps this was because the news story focused on the attack of the terrorists and not on the suffering of the victims. The focus on attacks may have directed the attention of audience members to feelings of threat instead of feelings of compassion for others, but this is speculation.

There is also a natural confound in our manipulation of exemplar strength. We selected images from the high strength video stimuli for use in the low strength still-images condition. Although we believe that this is a reasonable manipulation of exemplar strength, the selection of certain images naturally excludes other images. If included, these images could have feasibly reduced emotionality in our low strength condition. Although it is possible that this limitation caused our failure to find differences between the high and low exemplar strength conditions, we think the preceding recency explanation is more likely. It seems more probable that excluded images would have increased emotionality. As such, we feel that the lack of differences between our high and low exemplar strength conditions in emotionality was likely due to recency.

Finally, in addition to potential stimulus limitations, the self-report measure of intuition salience used in our study is a concern. The MFQ is a scale widely used to assess moral intuitions, which was our principal reason for its inclusion in the present study. As previously noted, however, the indicators used to measure ingroup loyalty do not include items assessing bias against outgroup members. Although this may have been an intentional decision by the scale developers, who label the measure "ingroup," this feature seems to ignore the outgroup distrust component of the broader conception of ingroup loyalty found in previous MFT literature (Haidt & Graham, 2007). Moreover, for our purposes, the scale limits its comparability to prior research examining outgroup bias (e.g., Das et al., 2009).

Given these limitations and the fact that our study is one of the first to test the mediation process outlined in MIME in the context of terrorist news the findings of this study have to be treated very cautiously. Future research is advised to replicate the study in order to further empirically underpin the MIME's theoretical predictions.

CONCLUSION

Our study attempted to examine the impact of a single exposure to terrorist attack news on the salience of moral intuitions and outgroup prejudice. Our findings demonstrated its ability to increase the salience of the authority intuition, which in turn decreased the willingness to help outgroup members. This observation raises important questions regarding the ability of terrorist attack news to impact the authority intuition and the mechanisms associated with it. Future research is needed not only to replicate the short-term findings of the current study but also to investigate the potential long-term processes. Is heavy exposure to news of terrorist attacks positively related to the chronic salience of the authority intuition? Is the chronic salience of the authority intuition negatively related to support for outgroup members, or other related outcomes not examined in the current study, such as voting behaviors, the support for extremist groups, or the committing of hate crimes?

We began our study by arguing that consideration of intuitive motivations could add to understandings of the way that exposure to terrorist news can influence explicit attitudes and social prejudice. Although extensive research examines media's impact on attitudes, less attention has been paid to fundamental underlying instincts upon which specific attitudes are based. The findings here suggest the potential value of the MIME for explicating mechanisms through which media exposure can influence tolerance for other people and ideas. However, this is only an initial attempt to apply the MIME to this area of research, and considerable work is needed before we can be confident in its value to understanding the impact of terrorist attack news.

REFERENCES

Alderman, L., & Ardley, J. (2015, November 13). Paris terror attacks leave awful realization: Another massacre. *The New York Times*. Retrieved from http://www.nytimes.com/2015/11/14/world/europe/paris-terror-attack.html?_r=0

Althaus, S. L. (2002). American news consumption during times of national crisis. *PS: Political Science & Politics, 35*, 517–521. doi:10.1017/S104909650200077X

Boomgaarden, H. G., & De Vreese, C. H. (2007). Dramatic real-world events and public opinion dynamics: Media coverage and its impact on public reactions to an assassination. *International Journal of Public Opinion Research, 19*, 354–366. doi:10.1093/ijpor/edm012

Bozzoli, C., & Müller, C. (2011). Perceptions and attitudes following a terrorist shock: Evidence from the UK. *European Journal of Political Economy, 27*, S89–S106. doi:10.1016/j.ejpoleco.2011.06.005

Burke, B. L., Martens, A., & Faucher, E. H. (2010). Two decades of terror management theory: A meta-analysis of mortality salience research. *Personality and Social Psychology Review, 14*, 155–195. doi:10.1177/1088868309352321

Byers, B. D., & Jones, J. A. (2007). The impact of the terrorist attacks of 9/11 on anti-Islamic hate crime. *Journal of Ethnicity in Criminal Justice, 5*, 43–56. doi:10.1300/J222v05n01_03

Das, E., Bushman, B. J., Bezemer, M. D., Kerkhof, P., & Vermeulen, I. E. (2009). How terrorism news reports increase prejudice against outgroups: A terror management account. *Journal of Experimental Social Psychology, 45*, 453–459. doi:10.1016/j.jesp.2008.12.001

Echebarria-Echabe, A., & Fernández-Guede, E. (2006). Effects of terrorism on attitudes and ideological orientation. *European Journal of Social Psychology, 36*, 259–265. doi:10.1002/ejsp.294

Florian, V., & Mikulincer, M. (1997). Fear of death and the judgment of social transgressions: A multidimensional test of terror management theory. *Journal of Personality and Social Psychology, 73*(2), 369–380. doi:10.1037/0022-3514.73.2.369

Graham, J., Haidt, J., & Nosek, B. A. (2009). Liberals and conservatives rely on different sets of moral foundations. *Journal of Personality and Social Psychology, 96*, 1029–1046. doi:10.1037/a0015141

Graham, J., Nosek, B. A., Haidt, J., Iyer, R., Koleva, S., & Ditto, P. H. (2011). Mapping the moral domain. *Journal of Personality and Social Psychology, 101*, 366–385. doi:10.1037/a0021847

Greenberg, J. (1987). The college sophomore as guinea pig: Setting the record straight. *The Academy of Management Review, 12,* 157–159. Retrieved from http://www.jstor.org/stable/258001

Greenberg, J., Pyszczynski, T., & Solomon, S. (1986). The causes and consequences of a need for self-esteem: A terror management theory. In R. F. Baumeister (Ed.), *Public and private self* (pp. 189–212). New York, NY: Springer-Verlag.

Grizzard, M., Plante, A. H., Huang, J., Weiss, J., Novotny, E., & Ngho, Z. (2015, November). *Graphic violence as moral motivator: Viewing atrocities makes us morally sensitive.* Paper presented at the Annual Conference of the National Communication Association, Las Vegas, NV.

Haidt, J., & Graham, J. (2007). When morality opposes justice: Conservatives have moral intuitions that liberals may not recognize. *Social Justice Research, 20,* 98–116. doi:10.1007/s11211-007-0034-z

Haidt, J., & Joseph, C. (2007). The moral mind: How five sets of innate intuitions guide the development of many culture specific virtues, and perhaps even modules. In P. Carruthers, S. Laurence, & S. P. Stich (Eds.), *Evolution and cognition. The innate mind* (pp. 367–392). New York, NY: Oxford University Press.

Hayes, A. F. (2013). *Introduction to mediation, moderation, and conditional process analysis: A regression-based approach.* New York, NY: Guilford Press.

Hetherington, M., & Suhay, E. (2011). Authoritarianism, threat, and americans' support for the war on terror. *American Journal of Political Science, 55,* 546–560. doi:10.1111/j.1540-5907.2011.00514.x

Kohlberg, L., Levine, C., & Hewer, A. (1983). *Moral stages: A current formulation and a response to critics; 9 tables. Contributions to human development* (Vol. 10). Basel, Switzerland: Karger.

Kugler, M., Jost, J. T., & Noorbaloochi, S. (2014). Another look at moral foundations theory: Do authoritarianism and social dominance orientation explain liberal-conservative differences in "moral" intuitions? *Social Justice Research, 27,* 413–431. doi:10.1007/s11211-014-0223-5

Leach, C. W., Van Zomeren, M., Zebel, S., Vliek, M. L. W., Pennekamp, S. F., Doosje, B., . . . Spears, R. (2008). Group-level self-definition and self investment: A hierarchical (multi-component) model of in-group Identification. *Journal of Personality and Social Psychology, 95,* 144–165.

McKinley, C. J., & Fahmy, S. (2011). Passing the "Breakfast Test": Exploring the effects of varying degrees of graphicness of war photography in the new media environment. *Visual Communication Quarterly, 18*(2), 70–83. doi:10.1080/15551393.2011.574060

Mogensen, K. (2008). Television journalism during terror attacks. *Media, War & Conflict, 1,* 31–49. doi:10.1177/1750635207087624

Simons, R. F., Detenber, B. H., Roedema, T. M., & Reiss, J. E. (1999). Emotion processing in three systems: The medium and the message. *Psychophysiology, 36,* 619–627.

Tamborini, R. (2011). Moral intuition and media entertainment. *Journal of Media Psychology: Theories, Methods, and Applications, 23,* 39–45. doi:10.1027/1864-1105/a000031

Tamborini, R. (2013). A model of intuitive morality and exemplars. In R. C. Tamborini (Ed.), *Media and the moral mind* (pp. 43–74). Abingdon, UK: Routledge.

Tamborini, R., Bowman, N., Prabhu, S., Hahn, L., Klebig, B., Grall, C., & Novotny, E. (2016). The effect of moral intuitions on decisions in video-game play: The role of temporary and chronic intuition accessibility. *New Media and Society.* Advance online publication. doi:10.1177/1461444816664356

Tamborini, R., Prabhu, S., Hahn, L., Idzik, P., & Wang, L. (2014, May). *News exposure's influence on the salience of moral intuitions: Testing the reliability of the Intuitive Motivation–Affect Misattribution Procedure (IM–AMP).* Paper presented at the 64th annual meeting of the International Communication Association, Seattle, WA.

Tamborini, R., Prabhu, S., Lewis, R. J., Grizzard, M., & Eden, A. (2016). The influence of media exposure on the accessibility of moral intuitions and associated affect. *Journal of Media Psychology.* doi:10.1027/1864-1105/a000183

Tamborini, R., Prabhu, S., Wang, L., & Grizzard, M. (2013, June). Setting the moral agenda: News exposure's influence on the salience of moral intuitions. In *63rd Annual Conference of the International Communication Association*, London, England.

Traugott, M., Brader, T., Coral, D., Curtin, R., Featherman, D., Groves, R., ... Willis, R. (2002). How Americans responded: A study of public reactions to 9/11/01. *PS: Political Science & Politics, 35*, 511–516. doi:10.1017/S1049096502000768

Van Leeuwen, F., & Park, J. H. (2009). Perceptions of social dangers, moral foundations, and political orientation. *Personality and Individual Differences, 47*, 169–173. doi:10.1016/j.paid.2009.02.017

Van Leeuwen, F., Park, J. H., Koenig, B. L., & Graham, J. (2012). Regional variation in pathogen prevalence predicts endorsement of group–focused moral concerns. *Evolution and Human Behavior, 33*, 429–437. doi:10.1016/j.evolhumbehav.2011.12.005

Zillmann, D. (2002). Exemplification theory of media influence. In J. Bryant & D. Zillmann (Eds.), *Media effects: Advances in theory and research* (2nd ed., pp. 19–41). Mahwah, NJ: Erlbaum.

Zillmann, D., & Brosius, H.-B. (2000). *Exemplification in communication: The influence of case reports on the perception of issues.* Mahwah, NJ: Erlbaum.

"Muslims are not Terrorists": Islamic State Coverage, Journalistic Differentiation Between Terrorism and Islam, Fear Reactions, and Attitudes Toward Muslims

Christian von Sikorski, Desirée Schmuck, Jörg Matthes, and Alice Binder

Previous research shows that news about Islamist terrorism can seriously affect citizens' fear reactions and influence non-Muslims' out-group perceptions of Muslims. We argue that news coverage that explicitly links Islam to terrorism or terrorists of the Islamic State (IS) may trigger fears in non-Muslim individuals. In contrast, news differentiation (i.e., explicitly distinguishing between Muslims and Muslim terrorists) may dampen particular

Christian von Sikorski (Ph.D., German Sport University Cologne, 2014), is a postdoctoral researcher in the Department of Communication at the University of Vienna as well as a visiting professor of media psychology in the Institute for Communication Psychology and Media Education at the University of Koblenz-Landau. His research interests include media effects, research methods, and online and visual communication with a special interest in terrorism news coverage and political scandals.

Desirée Schmuck (Ph.D., University of Vienna, 2017), is a postdoctoral researcher in the Department of Communication at the University of Vienna. Her research interests include the effects of political communication with a special interest in right-wing populism and terrorism news coverage.

Jörg Matthes (Ph.D., University of Zurich, 2007) is a professor of communication science in and the director of the Department of Communication at the University of Vienna. His research interests include advertising effects, the process of public opinion formation, news framing, and empirical methods.

Alice Binder (M.A., University of Vienna, 2016) is a doctoral candidate in the Department of Communication at the University of Vienna. Her research interests include political participation and children and media.

fear reactions in citizens. To test the specific effects of news differentiation, a controlled laboratory experiment was conducted. Results showed that undifferentiated news about IS terrorism increased participants' fear of terrorism and resulted in hostile perceptions toward Muslims in general. However, fear of terrorism only enhanced hostile attitudes toward Muslims for individuals with negative and moderately positive prior experiences with Muslims. For those with very positive experiences, no such relationship was found. Implications of these findings for journalism practice and intergroup relations in multi-cultural societies are discussed.

Global terrorism has been described as one of the most important and pressing issues of our time (Beck, 2002; Riedel, 2011). According to a recent Gallup poll, Americans currently name "terrorism" as the most important U.S. problem superseding the formerly most important topic: "the economy" (Riffkin, 2015). Especially, the rise of the so-called IS[1] has significantly affected the perception of terrorism around the world. Since the (self-) proclamation of the IS or the IS-Caliphate on June 29, 2014, the worldwide coverage on the IS has been dominating the media landscape (Satti, 2015; Zhang & Hellmueller, 2016). International news networks are extensively reporting on the vast number of atrocities, terrorist attacks, tortures, and rapes committed by proponents of the IS (Satti, 2015), which is classified as terrorist organization by the U.S. Department of State (2016) and many other international governments.

Previous research revealed that prejudices against Muslims are pervasive in Western societies (Strabac & Listhaug, 2008) and the general media coverage of Muslims is regularly framed in negative ways (Ahmed & Matthes, 2016; Bowe, Fahmy, & Matthes, 2015; Bowe, Fahmy, & Wanta, 2013; Powell, 2011; see also Matthes & Schmuck, 2017; Schmuck & Matthes, 2015, 2017). Especially, Muslims are regularly depicted as out-group extremists (Chuang & Roemer, 2013; Mahony, 2010; Sides & Gross, 2013) and are linked to terrorist acts in terrorism news (Ahmed & Matthes, 2016; Gerhards & Schäfer, 2014; Satti, 2015). In this context, a different line of research has demonstrated that exposure to news reports about terrorism frequently elicits fear of terrorism in news consumers (Nellis & Savage, 2012; Slone, 2000) and generates negative attitudes toward out-groups, as for instance Arabs (Das, Bushman, Bezemer, Kerkhof, & Vermeulen, 2009; Oswald, 2005; Stephan, Renfro, Esses, Stephan, & Martin, 2005).

Against this background, the intensive media coverage about the IS has raised concerns about unwanted and negative effects on news consumers' perceptions of Muslim citizens living in Western societies. More precisely, an intensive debate arose over whether there is a link between Islam and terrorists of Muslim faith (e.g.,

---

[1] According to the Stanford Mapping Militant Organizations Project (2016) the IS is a Salafi militant organization in Syria and Iraq that is also known as the Islamic State in Iraq and Syria (ISIS or ISIL). Throughout this article we use the name IS to refer to this organization.

Hodges, 2015), and how the news media should report about Islamist terrorists (Gerhards & Schäfer, 2014; Sides & Gross, 2013; Strabac & Listhaug, 2008). According to Arciszewski, Verlhiac, Goncalves, and Kruglanski (2010), news about "Islamist terrorism" especially affects Muslims living in European countries and the United States, because non-Muslim individuals may attribute fears and threats triggered by terrorism news to Muslims as a group.

We theorize that *differentiation* in news reports may play an essential role in preventing negative stereotypes about out-group members such as Muslims (Sides & Gross, 2013; Woods, 2011). Gerhards and Schäfer (2014) showed that the level of differentiation in terrorism news may significantly vary. For example, in the coverage of the 2005 terror attack in London, CNN tended to not differentiate between the Islamist terrorists and Muslims in general, describing the terrorists as "British-based radical Muslims" (Gerhards & Schäfer, 2014, p. 14). In contrast, the pan-Arab station Al Jazeera (Qatar) clearly distinguished between terrorists and Muslims in general, for example, "Many Muslims have condemned the attacks" (Gerhards & Schäfer, 2014, p. 12). Based on research on categorization (Hogg & Reid, 2006), journalistic differentiation may help news consumers to clearly differentiate between potentially overlapping out-group categories. This may, especially, apply to out-group categories that are not well differentiated in the first place (e.g., Muslims and Muslim extremists). More specifically, mass-mediated information about IS terrorism that does explicitly distinguish between Muslim terrorists and Muslims in general may result in a more differentiated perception of the out-group among non-Muslim recipients (i.e., Muslim terrorists are perceived as different from Muslims in general), preventing negative stereotypes toward Muslims. In contrast, undifferentiated news coverage (i.e., with no explicit differentiation between Muslim terrorists and the general Muslim population) may increase a non-Muslim recipient's perceived out-group homogeneity, thus increasing the perception that all Muslims are similar. This out-group homogeneity effect (Park & Rothbart, 1982; Rothgerber, 1997) describes the general tendency to assume that outgroup members are relatively more homogeneous than in-group memebers and has been demonstrated for different areas of research such as race (e.g., Aboud, 2003; Ryan, Judd, & Park, 1996) or nationality (e.g., Marques & Paez, 1994). Based on this body of research, we theorize that a lack of differentiation in news coverage about Muslim terrorists may result in more negative attitudes toward Muslim citizens in general (Oswald, 2005; Sides & Gross, 2013). However, this has not been tested so far.

This study aims to fill three important gaps. First, previous research lacks an investigation of how differentiation in news reports about Islamist terrorism affect recipients' perceptions of Muslims in general. Understanding the effects of differentiation can, however, be regarded as important insight for journalists in the newsroom. Second, fear of terrorism has been identified as an important mediator between terrorism news and particular out-group perceptions (e.g., Oswald, 2005; Slone, 2000). However, it remains unclear if and how differentiated, respectively, undifferentiated news reports about IS terrorism affect news consumers' fear reactions. Clarification of the under-lying processes of how differentiation in the news about terrorism affects news consumers' perceptions of Muslims can be regarded highly relevant for future research and may further contribute to improving news coverage about terrorism.

Third, certain individual predispositions such as personal experiences with Muslims may bolster or, by contrast, prevent negative effects of terrorism news on attitudes toward Muslims. Yet, the conditions that may moderate the extent to which news articles about terrorism influence individuals' attitudes toward Muslims have not been addressed thus far. However, as hostile attitudes toward minority groups may severely harm intergroup relations between Muslims and non-Muslims, investigating the conditions that make individuals particularly susceptible to negative effects of terrorism news reports is of crucial importance.

Therefore, the present research used an experimental design to manipu-late differentiation in the news coverage about IS terrorism. We presented different news articles dealing with IS terrorism to participants. The articles either differentiated between Muslim terrorists and the mainstream Muslim population or not. Hereby, we aimed at addressing the following three questions. First, does differentiation in news about IS terrorism affect recipients' fear of terrorism? Second, does fear of terrorism mediate the effects of article type (differentiated vs. undifferentiated) on negative atti-tudes toward Muslims? Third, does (positive/negative) prior personal experi-ence with Muslims moderate the effects of fear of terrorism on negative attitudes toward Muslims?

TERRORISM, NEWS COVERAGE, AND EFFECTS ON FEAR OF TERRORISM

Scholars have pointed out that defining terrorism can be regarded a difficult task and that what particular actions qualify as terrorist acts is a challenging question (for an overview, see Cooper, 2001). However, Moghaddam and Marsella (2004) suggested that several aspects can be pointed out that all terrorist acts have in common. They named the use of violence, the intention to generate fear, and the intention to influence or change citizens' political beliefs or social positions (p. 14). Furthermore, domestic terrorism (e.g., conflicts between a specific independence or separatist movement like the Irish Republican Army or Euskadi Ta Askatasuna) can be differentiated from international or global terrorism (that the current article focuses on), which is nowadays regularly "associated to radical Islamic fundamentalists calling for Jihad and to most people directly linked to Islam and Muslims" (Arciszewski et al., 2010, p. 7). Mass media play an essential role in distributing information about terrorism. Especially, after the self-proclamation of the IS in 2014, the international news media have started to intensively report about IS terrorist actions (Satti, 2015; Zhang & Hellmueller, 2016). In this context, it has been shown that the IS developed a systematic and elaborate media strategy that includes professionally produced media contents (e.g., articles, pictures, videos) that regularly appear on Western news media outlets (Farwell, 2014).

Previous research shows that news about terrorism—as intended by terrorist organizations like the IS (Cooper, 2001)—can seriously affect an individual's fear reactions (Lerner, Gonzalez, Small, & Fischhoff, 2003).[2] More precisely, news reports about terrorism can prime fear of terrorism in news consumers (Lerner et al., 2003). Priming refers to the "short-term impact of exposure to a mass-mediated stimulus on subsequent judgments" (Arendt, 2013) and particular primes may change "the standards that people use to make political evaluations" (Iyengar & Kinder, 1987, p. 63).

In general, Nellis and Savage (2012) showed that frequency of exposure to TV news about terrorism was associated with greater fear "that a family member will be the victim of a terrorist attack" (p. 7). Dumont, Yzerbyt, Wigboldus, and Gordijn (2003) showed that exposing European participants to visual material of the 9/11 attacks (Twin Towers on fire) and making them believe that they are part of the same group as the American victims of the attacks systematically increased their fear reactions. Slone (2000) demonstrated that participants exposed to a television news clip about terrorism in Israel (Hizballah and Hamas activities and terrorist attacks within Israel) experienced a significantly higher level of anxiety compared to a control group. Furthermore, Fischer, Greitemeyer, Kastenmüller, Frey, and Oßwald (2007) showed that the level of fear experienced by recipients depends on the degree of threat related information. Exposing participants to information about terrorist acts and highlighting a "currently high risk of terrorism" significantly increased recipients' fear reactions compared to the identical text in combination with information highlighting a current "low risk of terrorism."

Taken together, the reviewed results suggest that news about terrorism can systematically affect news consumers' fear reactions. However, the experienced level of fear of recipients may vary depending on the specific information provided or the way terrorism news are presented. We theorize that one characteristic of journalistic texts is crucial in that context: the level of journalistic differentiation between Muslim terrorists, on one hand, and Muslims in general, on the other.

---

[2] Fear has been described as an emotion that is negatively valenced and that is regularly accompanied by a high level of arousal. Fear is elicited by a threat that is perceived to be of high personal relevance to an individual (Lang, 1984).

NEWS DIFFERENTIATION: DISTINGUISHING BETWEEN MUSLIMS AND MUSLIM TERRORISTS

Previous research suggests that *differentiation* can play an essential role in news reports about terrorism (Gerhards & Schäfer, 2014; Sides & Gross, 2013; Stephan et al., 2005; Woods, 2011). Undifferentiated reporting, also referred to as generalization, "refers to the extension of the characteristics or activities of a specific and specifiable group of people to a much more general and open-ended set" (Teo, 2000, p. 16). Differentiated reporting, in contrast, explicitly distinguishes between more general and open-ended sets (e.g., Muslims in general) and the characteristics and activities of individual people (e.g., Islamist terrorists who self-identify as Muslims). Therefore, news differentiation is defined as the differentiation between Islamist terrorism or terrorist acts committed by Muslim terrorists *and* Muslims or the general Muslim population living in Western countries at large.

Non-Muslim recipients exposed to news coverage about terrorism that does not differentiate between Muslims and Muslim terrorists should increase a recipient's perception that all Muslims including Muslim terrorists are similar. This is in line with social identity theory (SIT; Tajfel & Turner, 1986), as well as the out-group homogeneity effect (Park & Rothbart, 1982; Rothgerber, 1997). SIT predicts that in-group individuals regularly try to maintain a positive social identity by comparisons with relevant out-groups. According to SIT, social behavior can range from *interpersonal* to *intergroup* behavior along a continuum (the extreme form of each behavior is only rarely found in real life). Interpersonal behavior relates to the "interaction between two or more individuals that is fully determined by their interpersonal relationships and individual characteristics, and not at all affected by various social groups or categories to which they respectively belong" (Tajfel & Turner, 1986, p. 277).

In contrast, the extreme form of intergroup behavior is defined as the interaction between two or more individuals (or groups of individuals) that is completely determined by their respective group memberships and not by their particular personal relationships (Tajfel & Turner, 1986). Especially, when social behavior can be characterized as intergroup behavior, in-group individuals tend to perceive out-group members as homogenous and similar (Tajfel & Turner, 1986).

Similarly, the out-group homogeneity effect (Park & Rothbart, 1982; Rothgerber, 1997) predicts that persons of an in-group (e.g., non-Muslims) will perceive an out-group (e.g., Muslims) as less variable and out-group members as more similar to one another, compared to the own in-group (e.g., "they" are all alike and we—in-group—are diverse). Thus, undifferentiated (compared to differentiated) terrorism news should increase a non-Muslim individual's perceived out-group homogeneity. Perceiving Muslim terrorists and Muslims as rather similar and as jointly being part of a rather large out-group (Muslim population in a Western society) may affect non-Muslim individuals' perceived fear of terrorism (Ommundsen, Van der Meer, Yakushko, & Ulleberg, 2013). The rationale behind this assumption is as follows: the larger an out-group technically is, the greater is the likelihood of being harmed by an out-group member. Rephrased, if all Muslims are perceived as terrorists, then the terrorist threat is much higher compared to the notion that Muslim terrorists are individual outsiders not associated with the Muslim majority. This is in line with previous research. According to Stephan, Ybarra, and Morrison (2009), relative group size of an out-group plays a fundamental role for intergroup perceptions. Following this, the larger an out-group is perceived to be, the higher the respective threat perceptions for an in-group member. Therefore, terrorism news that does not differentiate between Muslims in general (e.g., Muslim citizens living in a Western city) and individual Islamist terrorists should increase the perceived group size of the out-group, hence increasing non-Muslim individuals' fear of terrorism. In contrast, recipients exposed to differentiated news about terrorism should perceive the particular group size of the out-group to be smaller and will thus experience lower levels of fear of terrorism. Hence, we assume that both undifferentiated and differentiated news about IS terrorism will prime individuals' fear of terrorism. However, based on the theoretical reasoning just outlined, we expect that the former will have a stronger effect on fear of terrorism than the latter.

More formally, H1 and H2 read as follows:

H1: Participants exposed to (a) undifferentiated and (b) differentiated news about IS terrorism will experience an increased level of fear of terrorism compared to participants in the control group.

H2: Participants exposed to an undifferentiated news coverage about IS terrorism will experience a higher level of fear of terrorism compared to participants exposed to a differentiated news coverage.

EFFECTS OF FEAR OF TERRORISM ON HOSTILE ATTITUDES TOWARD MUSLIMS

According to intergroup threat theory (Stephan et al., 2009) perceptions of threat may significantly affect how individuals perceive an out-group or out-group members. Intergroup threat theory differentiates between two types of threats—*symbolic* and *realistic* threats (Stephan et al., 2009; for an earlier version of the theory labeled integrated threat theory, see Stephan & Stephan, 2000). Symbolic threats comprise threats to a group's values, beliefs, and worldviews, as well as individual threats such as undermining an individual's self-identity or self-esteem. Realistic threats are threats to a group's resources and general welfare, as well as threats to an individual like "actual physical or material harm, pain, torture, death, economic loss and threats to health or personal security" (Stephan et al., 2009, p. 44f.)

Empirical findings have supported the assumptions of the theory and shown that threat perceptions, respectively, fear of terrorism affects non-Muslim recipients' attitudes and hostile perceptions toward Muslims in general (Das et al., 2009; Lerner et al., 2003; Oswald, 2005; Stephan et al., 2005). Oswald (2005) showed that fear of terrorism is directly related to non-Muslim persons' hostile perceptions of Muslims in general, and Stephan et al. (2005) demonstrated that negative out-group perceptions were particularly strong when an out-group posed both realistic and symbolic threats.

One may argue that IS terrorism communicated via the news media poses both realistic and symbolic threats to news consumers and may thus result in particularly negative out-group perceptions. "Muslims in European countries (as well as in the United States) will be the first target" (p. 14) of such negative out-group perceptions because "fears and threats created by terrorism can be attributed to Muslims as a group" (Arciszewski et al., 2010, p. 7). Based on this reasoning, H3 reads as follows:

H3: A higher fear of terrorism will result in more hostile attitudes toward Muslims.

It has been proposed that intergroup contact (contact hypothesis) may improve in-group individuals' out-group attitudes (Pettigrew, 1998). Superficial and simple forms of contact tend to be ineffective (Hewstone & Brown, 1986) or may even result in more negative evaluations of the out-group, when respective contact is negative (Islam & Hewstone, 1993). However, more intensive positive experiences with out-group members (e.g., at the workplace) can improve attitudes toward the out-group at large (Brown, Vivian, & Hewstone, 1999; Voci & Hewstone, 2003), because individuals of an in-group "generalize positive attitudes promoted by the contact experience to include other members of the out-group not actually present in the contact situation" (Hewstone, 1996, p. 328).

The literature contains two prevalent models that make opposite predictions on how contact may/may not result in generalization, respectively, in positive effects toward out-group members not present in a contact situation. According to the decategorized contact model (DCM; Brewer & Miller, 1984) it is unlikely that contact to individuals of an out-group will result in a positive generalization to the out-group at large. The DCM predicts that positive contact of an in-group person will result in "decategorization" decreasing the salience of the out-group interactant's group membership while increasing the salience of his or her unique individual characteristics. Thus, a "non-category-based interaction" usually results in the perception that a particular person is an "exception to the rule" and that she or he is not a typical representative of the larger out-group (e.g., she or he is a nice person, but all other Muslims are not). In contrast, the mutual intergroup differentiation model (Hewstone, 1996; Hewstone & Brown, 1986) predicts that positive contact results in positive generalization when the original group membership of interactants' is preserved. Thus, it is important that in-group individuals have to be more or less aware of the out-group membership of an out-group interactant. In this context, Voci and Hewstone (2003, Experiment 1) showed that rather low levels of out-group salience were sufficient to generate respective positive out-group effects.

What remains unclear in this connection is, if positive prior contact with out-group individuals can mitigate negative out-group generalizations triggered by fear inducing media coverage. Previous research has mainly focused on the question of how positive contact between in-group and out-group individuals may or may not result in more positive attitudes toward the out-group at large. What has not been systematically studied in this context is, if positive real-life experiences with out-group members can alleviate potential hostile perceptions toward the out-group at large when negative media information is available. Islam and Hewstone (1993) showed that positive prior experiences (contact) were positively and particular negative prior experiences were negatively associated with attitude toward the out-group. However, what happens when in-group individuals with positive prior out-group experiences are confronted with threatening out-group media coverage? Can positive prior contact mitigate negative generalizations toward the out-group at large? Or does positive prior contact have no respective mitigating effects on out-group generalization? Based on this, we examine how positive prior experience moderates the particular influence of fear of terrorism on hostile perceptions toward Muslims with the help of the following research question.

RQ: How does positive prior experience with Muslims moderate the influence of fear of terrorism on hostile perceptions toward Muslims?

METHOD

We conducted a lab experiment with 103 participants.[3] We employed a between-subjects design with three experimental conditions. Participants were randomly assigned to the three conditions: three undifferentiated news articles about terrorism ($n = 35$), three otherwise identical differentiated news articles about terrorism ($n = 34$), and a control group that was exposed to three neutral news articles with no relation to the topic of terrorism ($n = 34$). A randomization check for age, $F(2) = .06$, ns; gender, $\chi^2 (2) = 2.94$, ns; immigration background, $\chi^2(4) = 4.71$, ns; and political predisposition, $F (2) = .03$, ns, was successful.

---

[3] The original sample consisted of 172 individuals. A second experimental factor had to be dropped from the analysis due to errors in the stimuli. In addition, we excluded one participant who indicated to be Muslim. This resulted in a total remaining sample of 103 participants.

Procedure

The experiment was conducted at the research laboratory of the Department of Communication at the University of Vienna, Austria. After prior informed consent, participants took part in the experiment in groups of a maximum of eight people. Each participant took a seat in front of one of the eight computers. During the study, participants were separated from each other by nontransparent, noise-absorbing dividers. They were randomly assigned to one of the stimulus conditions. In total, they read three newspaper articles, which were presented on a blank screen. Exposure time was not forced.[4] Participants were instructed to read the articles carefully. The stimulus presentation was followed by the assessment of the dependent variables. Upon completion, participants were thanked and debriefed.

Participants

Participants were 103 non-Muslim students (76% female) ages 19–39 ($M = 21.95$, $SD = 2.76$) enrolled in an introduction course on communication research at the University of Vienna. Of all participants, 63.1% had no immigrant background, 20.4% were second-generation immigrants, and 16.5% were first-generation immigrants; all participants were fluent in German (political orientation $M = 4.45$, $SD = 1.72$; 10-point scale from 1 [*extremely left*] to 10 [*extremely right*]). They received extra course credit for participation.

Stimulus Material

Based on existing news articles about Islamist terrorism, three news articles (described next) about attempted terrorist attacks or other activities by the IS were created for each condition. We used three articles in each condition because it has been argued that using different media exemplars increases the external validity of experimental media effects studies (Reeves, Yeykelis, & Cummings, 2016). The news articles were designed as online news articles of the web portal of the two biggest Austrian quality newspapers—*Der Standard* (derstandard.at) and *Die Presse* (diepresse.com)—and a large Austrian tabloid newspaper, *Kurier* (kurier.at). The articles were elaborately designed (layout, colors, newspaper logos, typography) to make them look like authentic news articles that had actually been published on the web portals of the three online newspapers. All articles had between 250 and 300 words. In addition to the stories' hard facts

---

[4] Participants could decide how long the exposure time to a news article would be. Overall time to read the news articles was entered into the model as covariate in an additional analysis. Reading time had no influence on fear of terror ($b = -0.03$, $SE = 0.12$, $p = .79$), or hostile attitudes toward Muslims ($b = 0.07$, $SE = 0.08$, $p = .37$). Also, the significant effects depicted in Table 1 did not change after exposure time was controlled.

(identical in each of the story versions), all articles' contained an expert's statement, which either stressed the difference between the mainstream Muslim population in Austria and Islamist terrorists or did not explicitly make this distinction (see the appendix).

The first article dealt with a planned terrorist attack by the IS in Vienna that could be prevented by arresting the terrorist. The article contained a statement by a fictitious jihadism expert who commented the events by either stressing the difference between a few jihadists and the mainstream Muslim population in the differentiated version or describing that more and more young Muslims from the general population are drawn into jihadism. The second article described a warning by Europol that rated the danger of a terrorist attack by the IS in Vienna as high. In addition, the article contained a comment by the director of Europol, who mentioned in the undifferentiated version that terrorism is a danger that has its roots in the center of the Muslim society in Western European countries such as Austria. In the differentiated version, he stated that Muslims in Austria are an important help for Austrian authorities in the war against terrorism. Finally, the third article dealt with the illegal smuggle of passports in Austria. The text described that the IS uses false passports to illegally send terrorists into the country. This article contained an additional statement by a fictitious Islam expert in which he warns of a generalization from Muslim extremists to peaceful Muslims who are not involved in these incidents in the differentiated version of the article. In the undifferentiated version, the expert mentions that the Islam religion and its organizations have a problem with violence that has to be dealt with. All other passages of the articles remained unchanged, and participants in the experimental conditions were thus exposed to the identical general information.

Participants in the control group were exposed to three unrelated articles with no cue to terrorism, Islam, or immigration. The three articles dealt with pregnancy, tourism, and school students' homework. The articles were original news articles published in the same three news outlets as the treatment articles. The articles were slightly changed to resemble the other articles in length and structure.

Pretest of Stimulus Material

A pretest of the stimulus material was conducted with an independent sample ($N = 52$, 58 % female, $M_{age} = 22.67$, $SD = 5.54$). All items were assessed on a Likert-type scale from 1 (*strongly disagree*) to 7 (*strongly agree*). Perceived differentiation of the news coverage was measured with three items: "The news articles clearly distinguish between Muslim terrorist and Muslims in general," "The news articles stress that Islam and Islamism need to be distinguished," and "The news articles emphasize that Muslims in general strongly oppose Islamist terrorism" (Cronbach's α = .65, M = 2.13,

SD = 0.06). Results clearly indicated that perceived differentiation was rated significantly higher in the differentiated news articles (*M* = 4.49, *SD* = 1.65) than in the undifferentiated news articles (*M* = 1.9, *SD* = 0.93), *F*(1, 50) = 49.31, *p* < .001, η^2 = .50. In addition, participants completed three items assessing the perceived quality of the articles. The news articles were rated as equally *credible* (undifferentiated *M* = 3.44, *SD* = 1.12, differentiated *M* = 3.52, *SD* = 0.71), *F*(1, 50) = 0.08, *p* = .774, η^2 = .00, and equally *comprehensible* (undifferentiated *M* = 4.15, *SD* = 1.39, differentiated *M* = 4.25, *SD* = 0.60), *F*(1, 50) = 0.12, *p* = .728, η^2 = .00. Furthermore, perceived *journalistic quality* did not significantly differ between the news articles (undifferentiated *M* = 3.26, *SD* = 1.12, differentiated *M* = 3.31, *SD* = 0.66), *F*(1, 50) = 0.03, *p* = .855, η^2 = .00. The ratings of all articles were thus close to the mean value.

Measures

All items were measured on a 7-point Likert-type scale from 1 (*strongly disagree*) to 7 (*strongly agree*). Fear of terrorism was gauged using three items based on Fischer and colleagues (2007) ("How high do you rate the probability that a terrorist attack may occur in Vienna within the next 12 months?" "How high do you rate the probability that you could become a victim of a terrorist attack?" "I fear that a terrorist attack could occur near me"; Cronbach's α = .75, *M* = 3.40, *SD* = 0.70). Hostile attitudes toward Muslims were assessed with seven items based on scales by Lee, Gibbons, Thompson, and Timani (2009) and Park, Felix, and Lee (2007) ("If possible, I would avoid going to places where Muslims would be," "If I could, I would avoid contact with Muslims," "If I could, I would live in a place where there were no Muslims," "Muslims should not be allowed to work in places where many Austrians gather such as airports," "Muslims lack the ability to think independently; they follow their leaders like sheep," "Muslims want to take over the world," "I believe that Muslims support violence against non-Muslims"; Cronbach's α = .89, *M* = 1.84, *SD* = 0.33)[5]. Analyses of the scale's factor structure supported a unidimensional scale.

We assessed both quantitative and qualitative aspects of personal real-life contact with Muslims based on Voci and Hewstone (2003). Frequency of personal contact with Muslims was assessed with the item "In general, how frequently are you personally in contact with Muslims?" from 1 (*never*) to 7 (*very frequently*). Quality of prior contact experiences was gauged with the item "How would you rate your personal contact with Muslims thus far?" from 1 (*very negative*) to 7 (*very positive*).

---

[5] The scale represents a very conservative assessment of negative attitudes toward Muslims. Items' mean values and standard deviations in this study are comparable to those of the original scale (see Lee et al., 2009).

Data Analysis

To test our hypotheses and find an answer to our research question, we conducted a moderated mediation model using the PROCESS macro in SPSS. Experimental condition was dummy coded with the control group as reference group. Fear of terrorism was modeled as mediator of the newspaper articles' effects on hostile attitudes toward Muslims. Prior experience with Muslims was modeled as moderator (Figure 1 shows the full theoretical model). The interaction term between fear of terrorism and prior experience with Muslims was modeled by including their multiplicative term. Fear of terrorism and prior experiences with Muslims were mean centered prior to computing the product (Hayes, 2013). The 95% bias-corrected bootstrap confidence intervals based on 10,000 bootstrap samples were used for statistical inference of indirect effects. The frequency of contact with Muslims was controlled in all analyses to ensure that the quality of contact with Muslims was assessed independently of the contact frequency.

RESULTS

First, we investigated the effects of differentiation in news coverage about the IS on individuals' fear of terrorism (H1). Results revealed a positive and significant effect of the undifferentiated news articles on fear of terrorism compared to the control group ($b = 0.68$, $SE = 0.32$, $p = .035$). The differentiated news articles, in contrast, had no significant effect on fear of terrorism ($b = 0.47$, $SE = 0.32$, $p = .141$; see Table 1). Thus, news coverage that clearly differentiates between Muslim terrorists and Muslims in general does not enhance individuals' general fear of terrorism. In contrast, exposure to articles that lack such a differentiation

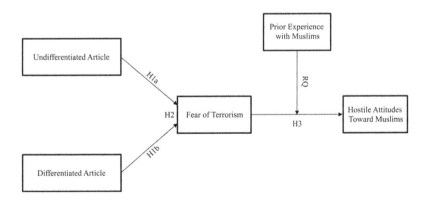

FIGURE 1 Theoretical model.

activates higher fear of terrorism (see Table 2 for mean values). This supports H1a and contradicts H1b.

Our second hypothesis postulated that undifferentiated news articles would exert a stronger effect on fear of terrorism than the effects of the differentiated news articles. To test this hypothesis we recoded our dummy variables with differentiated news articles as reference group and repeated the analysis (not shown here). Findings revealed that although the effect of the undifferentiated articles on fear of terrorism was stronger than the one of the differentiated articles, this difference lacked statistical significance ($b = 0.20$, $SE = 0.32$, $p = .682$). Thus, H2 has to be rejected. That means, although we—by trend—

TABLE 1
Ordinary Least Squares Path Analysis, Unstandardized Coefficients

Variables	Fear of Terrorism			Hostile Attitudes Toward Muslims		
	b	SE	β	b	SE	β
Undifferentiated news coverage	0.68	0.32	.24*	0.18	0.22	.08
Differentiated news coverage	0.47	0.32	.17	0.09	0.22	.04
Frequency of contact with Muslims	0.00	0.08	.00	0.04	0.06	.07
Fear of terrorism				0.19	0.07	.24**
Prior experience with Muslims				−0.39	0.07	−.46***
Fear of terrorism × Prior experience with Muslims				−0.10	0.05	−.16†
Adj. R^2	.05			.34		
F for change in Adj. R^2	1.61			8.28***		

Note. $N = 103$.
†$p < .10$. *$p < .05$. **$p < .01$. ***$p < .001$.

TABLE 2
Mean Values of Dependent Variables for Different Experimental Conditions

Variable	Control[a]	Undifferentiated News Articles[b]	Differentiated News Articles[c]
	M (SD)	M (SD)	M (SD)
Fear of terrorism	3.01 (1.26)	3.69 (1.37)	3.48 (1.27)
Hostile attitudes toward Muslims	1.61 (0.89)	1.99 (1.26)	1.97 (1.00)

Note. $N = 103$.
[a]$n = 34$. [b]$n = 35$. [c]$n = 32$.

observed a respective difference between the differentiated and undifferentiated article, this difference was not statistically significant.

Next, we investigated whether fear of terrorism increased hostile attitudes toward Muslims (H3). This was the case. The findings showed that fear of terrorism positively and significantly predicted hostile attitudes toward Muslims ($b = 0.19$, $SE = 0.07$, $p < .001$), which supports H3 (see Table 1). Furthermore, we tested whether this relationship was moderated by prior experiences with Muslims. We found a close-to-significant moderation effect of prior experience with Muslims on the effect of fear of terrorism on hostile attitudes toward Muslims ($b = -0.10$, $SE = 0.05$, $p = .056$). Using Moderated Mediation analysis (Model 14 in PROCESS), we investigated the indirect effect of the undifferentiated news article on hostile attitudes toward Muslims via fear of terrorism at different levels of the moderator (the mean and ± 1 SD from the mean). Findings revealed that fear of terrorism mediated the effects of the undifferentiated news coverage on hostile attitudes toward Muslims for individuals with less positive experiences toward Muslims (indirect effect of exposure: $b = 0.22$, $SE = 0.13$), confidence interval (CI) [.02, .56], and moderately positive experiences (indirect effect of exposure: $b = 0.13$, $SE = 0.08$), CI [.01, .34]. However, for individuals with very positive prior experiences with Muslims, we found no significant mediation effect of fear of terrorism on hostile attitudes toward Muslims (indirect effect of exposure: $b = 0.04$, $SE = 0.06$, CI [–.04, .25]. These results indicate that extremely positive encounters with Muslims prevent the transfer of fear of terrorism on hostile attitudes toward Muslims (Index of Moderated Mediation = -0.07, $SE = 0.05$), CI [–.21, –.00]. Hence, answering our research question (How does positive prior experience with Muslims moderate the influence of fear of terrorism on hostile perceptions toward Muslims?), we find that fear of terrorism only enhanced hostile attitudes toward Muslims for individuals with less positive and moderately positive prior experiences with Muslims. For those with very positive experiences, no such relationship was found (see Figure 2).

Overall, we found no direct effects of the undifferentiated news articles ($b = 0.18$, $SE = 0.22$, $p = .430$) or the differentiated news articles ($b = 0.09$, $SE = 0.22$, $p = .70$) on hostile attitudes toward Muslims. Prior experience with Muslims had a strong negative direct effect on hostile attitudes toward Muslims ($b = -0.39$, $SE = 0.07$, $p < .001$). Finally, frequency of contact with Muslims had no effect on fear of terror ($b = 0.00$, $SE = 0.08$, $p = .964$) or hostile attitudes toward Muslims ($b = 0.04$, $SE = 0.06$, $p = .446$). In total, the predictors explained 5% of the variance of fear of terror and 34% of the variance of hostile attitudes toward Muslims (see Table 1).

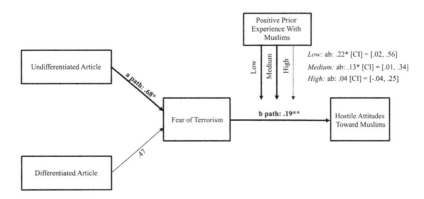

FIGURE 2 Moderated mediation model showing the indirect effects of the undifferentiated article on hostile attitudes toward Muslims via fear of terrorism for individuals with low, medium, and high positive prior experiences with Muslims. Unstandardized beta coefficients are shown, 95% bias-corrected bootstrap confidence intervals based on 10,000 bootstrap samples are shown for indirect effects. Bold lines indicate significant effects. $*p < .05. **p < .01. ***p < .001.$

DISCUSSION

This study examined the role of differentiation in news articles about IS terrorism. In line with intergroup threat theory (Stephan et al., 2009), the results revealed that undifferentiated news about IS terrorism indirectly affected participants' attitudes toward the general Muslim population in a negative way. Fear of terrorism operated as a mediator. That is, participants exposed to undifferentiated news experienced an increased level of fear of terrorism, and these fear reactions negatively influenced a participant's attitudes toward Muslims. In contrast, this relationship between exposure and hostile attitudes toward Muslims could not be detected for individuals exposed to differentiated news about IS terrorism.

Therefore, the results of the present study corroborate and extend earlier findings on the effects of fear of terrorism showing that terrorism news increase news consumers' fear reactions (e.g., Das et al., 2009; Oswald, 2005; Slone, 2000; Stephan et al., 2005). However, in these previous studies participants were assigned to specific terrorism stimuli (fear evoking news) or a control condition (exposure to an unrelated and non-fear-evoking topic) in order to examine particular effects of terrorism news on a superordinate level. The present study extended this line of research and did not only examine effects of terrorism news per se. Particular effects were analyzed in a more nuanced way showing that terrorism news does not necessarily have to increase a recipient's fear of terrorism, and that differentiation within the news played a crucial role in this context. In line with this reasoning, another central finding of the study relates to participants' prior experiences with Muslims. We found no significant mediation effect of fear of terrorism on hostile attitudes toward Muslims for participants who reported very positive prior experiences with Muslims. However, fear of terrorism enhanced hostile attitudes toward Muslims for both individuals with less positive and moderately positive prior experiences with Muslims.

These results indicate that hostile attitudes toward Muslims in connection with terrorism news can be prevented in two ways. First, although differentiated news reporting can be regarded relevant in general, it apparently is a necessary prerequisite for individuals with less positive or moderately positive prior experiences with Muslims. For those individuals, differentiation has to be realized within the news. Second, very positive prior experiences may avoid hostile attitudes toward Muslims in response to undifferentiated news. Individuals with very positive prior experiences with Muslims are obviously capable of differentiating between Muslims and Muslim terrorists—in form of a mental process—even when they are exposed to undifferentiated news about terrorism. That is, because they made positive prior experiences with Muslims, they are immune against generalizing terror news to all Muslims. This extends previous research on categorization and out-group attitudes (Hewstone, 1996; Hewstone & Brown, 1986; Islam & Hewstone, 1993) showing that positive out-group contact may not only result in positive generalizations toward the out-group at large but also buffer against negative media effects on hostile out-group attitudes. Furthermore, this finding underlines the importance of news differentiation in the context of terrorism news, because non-Muslim individuals' views of Muslims are generally rather negative (Strabac & Listhaug, 2008) and a large share of non-Muslims is not in regular communication with Muslims, thus decreasing the likelihood of very positive real-life experiences (Kalkan, Layman, & Uslaner, 2009).

Furthermore, our data clearly show that positive prior experiences could not prevent the activation of fear of terrorism in response to the undifferentiated news article. An additional analysis revealed no significant interaction effect between news differentiation and prior experiences on fear of terrorism ($b = -.02$, $SE = .22$, $p = .926$). Thus, this finding additionally confirms that very positive experiences can only prevent the transfer of fear of terrorism on attitudes toward Muslims. Fear in response to the undifferentiated article, however, is evoked among all individuals irrespectively of their prior experiences. Prior experiences can thus help to cope with fear, rather than preventing fear in the first place.

It is important to note that the two news conditions did not differ significantly from each other. Yet the undifferentiated condition had a significant effect compared to the control group, the differentiated condition did not. That is, although there were differences in statistical significance compared to the control group, this difference itself was not significant (although clearly pointing into the expected direction). One explanation for this could be the rather conservative manipulation used in the undifferentiated versions of our stimulus articles. Although participants were exposed to three articles, each article consisted of only one modified statement at the end of the particular article (see the appendix), whereas the rest of the information remained unchanged. However, news articles may contain numerous and much stronger hints or cues to potential links between Islam and radical Muslims. For example, the central theme of an article can deal with this aspect, and these links may be incorporated into a news article much more prominently (in the headline of a news article, e.g., "There is a clear link between Islam and terrorism"; see Hodges, 2015). Thus, one may argue that the detected effects in our study were realized under rather conservative conditions and one may expect even stronger effects in "real life." Furthermore, the effects of frequent exposure to undifferentiated (differentiated) news may add up, thus resulting in powerful effects across time (Abelson, 1985). Irrespective of this explanation, we can still interpret the statistically significant effect of the undifferentiated condition on fear, and we can conclude that differentiated news does not significantly evoke fear. This is in line with our theoretical reasoning.

There are notable limitations. First, future studies should try to replicate the present findings outside a forced exposure setting (see de Vreese & Neijens, 2016) and using different samples (e.g., older and less educated participants) and larger sample sizes. Second, participants in the present study were exposed to online news articles. Although we used several articles, as well as both quality and tabloid online newspapers as news outlets, future studies may use different news outlets (e.g., television), thus testing if the effects can be generalized. One may argue that a combination of textual/verbal and visual information may result in even stronger effects, because differentiated or undifferentiated information can be depicted more vividly using, for example, text and visuals. However, this is an empirical question and should be analyzed in future approaches. Third, we tested effects of differentiation on participants' fear reactions and respective out-group perceptions only. However, one may argue that reporting about terrorism in an undifferentiated/differentiated way may affect several other outcome variables including an individual's perceived news credibility. The rationale behind this assumption is that recipients may regard high levels of differentiation (compared to a low level of differentiation) depicted in a news article as a quality indicator increasing the perceived credibility of a news coverage or a news outlet. We can clearly rule out this explanation in our study: The pretest demonstrated that both versions were rated as equally credible. But future examinations should additionally test this theoretically important assumption. Fourth, future studies should examine the moderating effects of positive prior contact with Muslims in depth. Especially, examining the role of group salience, for example, how salient does a respective out-group membership have to be during interaction to result in positive generalizations toward the out-group, and how specific prior experiences interact with particular information provided by the news media would be valuable new avenues for future research.

CONCLUSION

The news media frequently and intensively report about terrorism news including attacks or attempted attacks by IS terrorists (Satti, 2015; Zhang & Hellmueller, 2016). Of course, news reports of this kind are valuable and important information for the public's understanding of certain issues at stake and can be regarded a cornerstone for free and democratic decision making. However, this type of news coverage can come with serious consequences, namely, negative out-group perceptions of Muslims in general and may further enhance intergroup conflicts between Muslims and non-Muslims. Differentiated news reporting may dissolve this dilemma. When journalists in the newsroom clearly and explicitly distinguish between news about Muslims and Islamist terrorism, this may result in a win-win situation. Citizens are being informed about serious issues at stake, whereas unwanted generalizations resulting in hostile attitudes toward Muslims are avoided.

REFERENCES

Abelson, R. P. (1985). A variance explanation paradox: When a little is a lot. *Psychological Bulletin*, *97*(1), 129–133. doi:10.1037/0033-2909.97.1.129

Aboud, F. E. (2003). The formation of in-group favoritism and out-group prejudice in young children: Are they distinct attitudes? *Developmental Psychology*, *39*(1), 48–60. doi:10.1037/0012-1649.39.1.48

Ahmed, S., & Matthes, J. (2016). Media representation of Muslims and Islam from 2000 to 2015: A meta-analysis. *International Communication Gazette*, *79*(3), 219–244. doi:10.1177/1748048516656305

Arciszewski, T., Verlhiac, J.-F., Goncalves, I., & Kruglanski, A. (2010). From psychology of terrorists to psychology of terrorism. *Revue Internationale de Psychologie Sociale*, *22*(3), 5–34. Retrieved from http://www.cairn.info/article.php?ID_ARTICLE=RIPSO_223_0005

Arendt, F. (2013). Dose-dependent media priming effects of stereotypic newspaper articles on implicit and explicit stereotypes. *Journal of Communication*, *63*(5), 830–851. doi:10.1111/jcom.12056

Beck, U. (2002). The terrorist threat. World risk society revisited. *Theory, Culture & Society*, *19*(4), 39–55. doi:10.1177/0263276402019004003

Bowe, B. J., Fahmy, S., & Matthes, J. (2015). U.S. newspapers provide nuanced picture of Islam. *Newspaper Research Journal*, *36*(1), 42–57. doi:10.1177/073953291503600104

Bowe, B. J., Fahmy, S., & Wanta, W. (2013). Missing religion: Second level agenda setting and Islam in American newspapers. *International Communication Gazette, 75*(7), 636–652. doi:10.1177/1748048513482544

Brewer, M. B., & Miller, N. (1984). Beyond the contact hypothesis: Theoretical perspectives on desegregation. In N. Miller & M. B. Brewer (Eds.), *Groups in contact: The psychology of desegregation* (pp. 281–302). Orlando, FL: Academic Press.

Brown, R., Vivian, J., & Hewstone, M. (1999). Changing attitudes through intergroup contact: The effects of group membership salience. *European Journal of Social Psychology, 29*(5–6), 741–764. doi:10.1002/(SICI)1099-0992(199908/09)29:5/6<741::AID-EJSP972>3.0.CO;2-8

Chuang, A., & Roemer, R. C. (2013). The immigrant Muslim American at the boundary of insider and outsider: Representations of Faisal Shahzad as "homegrown" terrorist. *Journalism & Mass Communication Quarterly, 90*(1), 89–107. doi:10.1177/1077699012468740

Cooper, H. H. A. (2001). Terrorism: The problem of definition revisited. *American Behavioral Scientist, 44*(6), 881–893. doi:10.1177/00027640121956575

Das, E., Bushman, B. J., Bezemer, M. D., Kerkhof, P., & Vermeulen, I. E. (2009). How terrorism news reports increase prejudice against outgroups: A terror management account. *Journal of Experimental Social Psychology, 45*(3), 453–459. doi:10.1016/j.jesp.2008.12.001

de Vreese, C. H., & Neijens, P. C. (2016). Measuring media exposure in a changing communications environment. *Communication Methods and Measures, 10*(2-3), 69–80. doi: 10.1080/19312458.2016.1150441

Dumont, M., Yzerbyt, V., Wigboldus, D., & Gordijn, E. H. (2003). Social categorization and fear reactions to the September 11th terrorist attacks. *Personality and Social Psychology Bulletin, 29*(12), 1509–1520. doi:10.1177/0146167203256923

Farwell, J. P. (2014). The Media Strategy of ISIS. *Survival, 56*(6), 49–55. doi:10.1080/00396338.2014.985436

Fischer, P., Greitemeyer, T., Kastenmüller, A., Frey, D., & Oßwald, S. (2007). Terror salience and punishment: Does terror salience induce threat to social order? *Journal of Experimental Psychology, 43*(6), 946–971. doi:10.1016/j.jesp.2006.10.004

Gerhards, J., & Schäfer, M. S. (2014). International terrorism, domestic coverage? How terrorist attacks are presented in the news of CNN, Al Jazeera, the BBC, and ARD. *International Communication Gazette, 76*(1), 3–26. doi:10.1177/1748048513504158

Hayes, A. F. (2013). *Introduction to mediation, moderation, and conditional process analysis: A regression-based approach.* New York, NY: Guilford Press.

Hewstone, M. (1996). Contact and categorization: Social psychological interventions to change intergroup relations. In C. N. Macrae, C. Stangor, & M. Hewstone (Eds.), *Stereotypes and stereotyping* (pp. 323–368). New York, NY: Guilford Press.

Hewstone, M., & Brown, R. J. (1986). Contact is not enough: An intergroup perspective on the contact hypothesis. In M. Hewstone, & R. J. Brown (Eds.), *Contact and conflict in intergroup encounters* (pp. 1–44). Oxford, UK: Blackwell.

Hodges, D. (2015). *There is a clear link between Islam and terrorism. It's up to all of us to break it.* Retrieved from http://www.telegraph.co.uk/news/uknews/terrorism-in-the-uk/12007852/There-is-a-clear-link-between-Islam-and-terrorism.-Its-up-to-all-of-us-to-break-it.html

Hogg, M. A., & Reid, S. (2006). Social identity, self-categorization, and the communication of group norms. *Communication Theory, 16*(1), 7–30. doi:10.1111/j.1468-2885.2006.00003.x

Islam, M. R., & Hewstone, M. (1993). Dimensions of contact as predictors of intergroup anxiety, perceived outgroup variability, and outgroup attitude: An integrative model. *Personality & Social Psychology Bulletin, 19*(6), 700–710. doi:10.1177/0146167293196005

Iyengar, S., & Kinder, D. R. (1987). *News that matters: Television and American opinion.* Chicago, IL: Chicago University Press.

Kalkan, K. O., Layman, G. C., & Uslaner, E. M. (2009). "Bands of others"? Attitudes toward Muslims in contemporary American Society. *Journal of Politics, 71*(3), 847–862. doi:10.1017/S0022381609090756

Lang, P. J. (1984). Cognition in emotion: Concept and action. In C. E. Izard, J. Kagan, & R. B. Zajonc (Eds.), *Emotions, cognition, and behavior* (pp. 192–226). Cambridge, UK: Cambridge University Press.

Lee, S. A., Gibbons, J. A., Thompson, J. M., & Timani, H. S. (2009). The Islamophobia Scale: Instrument development and initial validation. *The International Journal for the Psychology of Religion, 19*(2), 92–105. doi:10.1080/10508610802711137

Lerner, J. S., Gonzalez, R. M., Small, D. A., & Fischhoff, B. (2003). Effects of fear and anger on perceived risk of terrorism. *Psychological Science, 14*(2), 144–150. doi:10.1111/1467-9280.01433

Mahony, I. (2010). Diverging frames: A comparison of Indonesian and Australian press portrayals of terrorism and Islamic groups in Indonesia. *International Communication Gazette, 72*(8), 739–758. doi:10.1177/1748048510380813

Marques, J. M., & Paez, D. (1994). The 'black sheep effect': Social categorization, rejection of ingroup deviates, and perception of group variability. *European Review of Social Psychology, 5*(1), 37–68. doi:10.1080/14792779543000011

Matthes, J., & Schmuck, D. (2017). The effects of anti-immigrant right-wing populist ads on implicit and explicit attitudes: A moderated mediation model. *Communication Research, 44*(4), 556–581. doi:10.1177/0093650215577859

Moghaddam, F. M., & Marsella, A. J. (Eds.). (2004). *Understanding terrorism: Psychological roots, consequences, and interventions.* Washington, DC: American Psychological Association.

Nellis, A. M., & Savage, J. (2012). Does watching the news affect fear of terrorism? The importance of media exposure on terrorism fear. *Crime & Delinquency, 58*(5), 748–768. doi:10.1177/0011128712452961

Ommundsen, R., Van der Meer, K., Yakushko, O., & Ulleberg, P. (2013). Exploring the relationship between fear-related xenophobia, perceptions of out-group entitativity, and social contact in Norway. *Psychological Reports, 112*(1), 109–124. doi:10.2466/17.07.21.PR0.112.1.109-124

Oswald, D. L. (2005). Understanding anti-Arab reactions post 9/11: The role of threats, social categories, and personal ideologies. *Journal of Applied Social Psychology, 35*(9), 1775–1799. doi:10.1111/j.1559-1816.2005.tb02195.x

Park, B., & Rothbart, M. (1982). Perception of out-group homogeneity and levels of social categorization: Memory for the subordinate attributes of in-group and out-group members. *Journal of Personality and Social Psychology, 42*(6), 1051–1068. doi:10.1037/0022-3514.42.6.1051

Park, J., Felix, K., & Lee, G. (2007). Implicit attitudes toward Arab-Muslims and the moderating effects of social information. *Basic and Applied Social Psychology, 29*(1), 35–45. doi:10.1080/01973530701330942

Pettigrew, T. F. (1998). Intergroup contact theory. *Annual Review of Psychology, 49*, 65–85. doi:10.1146/annurev.psych.49.1.65

Powell, K. A. (2011). Framing Islam: An analysis of US media coverage of terrorism since 9/11. *Communication Studies, 62*(1), 90–112. doi:10.1080/10510974.2011.533599

Reeves, B., Yeykelis, L., & Cummings, J. J. (2016). The use of media in media psychology. *Media Psychology, 19*(1), 49–71. doi:10.1080/15213269.2015.1030083

Riedel, B. (2011, December 9). *The grave new world: Terrorism in the 21st century.* Retrieved from https://www.brookings.edu/articles/the-grave-new-world-terrorism-in-the-21st-century/

Riffkin, R. (2015, December 14). *Americans name terrorism as No. 1 U.S. problem.* Retrieved from http://www.gallup.com/poll/187655/americans-name-terrorism-no-problem.aspx

Rothgerber, H. (1997). External intergroup threat as an antecedent to perceptions of in-group and out-group homogeneity. *Journal of Personality and Social Psychology, 73*(6), 1206–1212. doi:10.1037/0022-3514.73.6.1191

Ryan, C. S., Judd, C. M., & Park, B. (1996). Effects of racial stereotypes on judgments of individuals: The moderating role of perceived group variability. *Journal of Experimental Social Psychology, 32* (1), 71–103. doi:10.1006/jesp.1996.0004

Satti, M. (2015). Framing the IS on Al Jazeera English and the BBC websites. *Journal of Arab & Muslim Media Research, 8*(1), 37–53. doi:10.1386/jammr.8.1.37_1

Schmuck, D., & Matthes, J. (2015). How anti-immigrant right-wing populist advertisements affect young voters: Symbolic threats, economic threats, and the moderating role of education. *Journal of Ethnic and Migration Studies, 41*(10), 1577–1599. doi:10.1080/1369183X.2014.981513

Schmuck, D., & Matthes, J. (2017). Effects of economic and symbolic threat appeals in right-wing populist advertising on anti-immigrant attitudes: The impact of textual and visual appeals. *Political Communication.* Advance online publication. doi:10.1080/10584609.2017.1316807

Sides, J., & Gross, K. (2013). Stereotypes of Muslims and support for the war on terror. *The Journal of Politics, 75*(3), 583–598. doi:10.1017/S0022381613000388

Slone, M. (2000). Responses to media coverage of terror. *Journal of Conflict Resolution, 44*(4), 508–522. doi:10.1177/0022002700044004005

Stanford Mapping Militant Organizations Project. (2016, April 4). Mapping militant organizations. *The Islamic State.* Retrieved from http://web.stanford.edu/group/mappingmilitants/cgi-bin/groups/view/1

Stephan, W. G., Renfro, C. L., Esses, V. M., Stephan, C. W., & Martin, T. (2005). The effects of feeling threatened on attitudes toward immigrants. *International Journal of Intercultural Relations, 29*(1), 1–19. doi:10.1016/j.ijintrel.2005.04.011

Stephan, W. G., & Stephan, C. W. (2000). An integrated threat theory of prejudice. In S. Oskamp (Ed.), *Reducing prejudice and discrimination* (pp. 23–45). Mahwah, NJ: Erlbaum.

Stephan, W. G., Ybarra, O., & Morrison, K. R. (2009). Intergroup threat theory. In T. D. Nelson (Ed.), *Handbook of prejudice, stereotyping, and discrimination* (pp. 43–59). New York, NY: Psychology Press.

Strabac, Z., & Listhaug, O. (2008). Anti-Muslim prejudice in Europe: A multilevel analysis of survey data from 30 countries. *Social Science Research, 37*(1), 268–286. doi:10.1016/j.ssresearch.2007.02.004

Tajfel, H., & Turner, J. C. (1986). The social identity theory of intergroup behavior. In S. Worchel & L. W. Austin (Eds.), *Psychology of intergroup relations* (pp. 276–293). Chicago, IL: Nelson-Hall.

Teo, P. (2000). Racism in the news: A critical discourse analysis of news reporting in two Australian newspapers. *Discourse & Society, 11*(1), 7–49. doi:10.1177/0957926500011001002

U.S. Department of State. (2016). *Foreign terrorist organizations.* Retrieved from http://www.state.gov/j/ct/rls/other/des/123085.htm

Voci, A., & Hewstone, M. (2003). Intergroup contact and prejudice toward immigrants in Italy: The mediational role of anxiety and the moderational role of group salience. *Group Processes & Intergroup Relations, 6*(1), 37–54. doi:10.1177/1368430203006001011

Woods, J. (2011). Framing terror: An experimental framing effects study of the perceived threat of terrorism. *Critical Studies on Terrorism, 4*(2), 199–217. doi:10.1080/17539153.2011.586205

Zhang, X., & Hellmueller, L. (2016). Transnational media coverage of the ISIS threat: A global perspective? *International Journal of Communication, 10,* 766–785.

APPENDIX

Excerpt of News Article 1

News Outlet: Der Standard (derstandard.at)

Headline: Extremism in Austria. Member of the IS planned terrorist attack in Vienna.
Statement: Timothy Welkner, jihadism expert of the University of Oxford (fictitious)
Undifferentiated version: "The Internet facilitates the networking and radicalization process of Muslims. Recently, young men and women out of the center of the Muslim society increasingly deal with the Islam religion. They visit the services at the Mosque, read Koran extracts or watch propaganda videos. In doing so, they sympathize with violent extremists or become more extremist themselves within a very short period of time."
Differentiated version: "The Internet facilitates the networking and radicalization of these people. They become extremists. At this point it is important to stress, however, that this propensity to violence has nothing to do with the Islam religion and the high number of well-integrated Muslims in our society. It's important to clearly distinguish between the IS and peaceful Muslims. These violent acts are incompatible with the Islam religion."

Excerpt of News Article 2

News Outlet: Kurier (kurier.at)

Headline: Europol warns of major terrorist attacks by the IS: High risk for Vienna
Statement: Rob Wainwright, Europol director
Undifferentiated version: "Self-radicalizations of Muslims often occur in the center of Muslim communities or mosques. In particular, Salafist tendencies provide the ground for these radicalizations. It's important that we locate Muslim extremists as soon as possible. This is a core task of the Europol anti-terror center".

Differentiated version: "In the war against terror by the IS Muslim organizations and communities provided crucial information. They are on our side. Together we pursue the same goal within the war against the terror by the IS. Without the help of our Muslim partners, we would not have been able to prevent several attacks by IS terrorists."

Excerpt of News Article 3

News Outlet: Die Presse (diepresse.com)

Headline: "Islamic State": Smuggled Austrian passports in Vienna
Statement: Hamid Hamdan, Islam expert

Undifferentiated version: "This is a serious problem. But an even bigger problem is the radicalization of young men who are born here. They already have Austrian passports. We have to take actions against these fundamentalists and extremist organizations, because Islam and its organizations naturally have a problem with violence."

Differentiated version: "This is a serious problem. But it's also a big problem that this could lead to general suspicion toward peaceful Muslims who have lived and worked here for a long time and have nothing to do with this kind of extremism and terror. We have to avoid this by all means. Christians, Jews and Muslims have to stand together and oppose these fundamentalists."

The full versions of all articles are not printed here due to space constraints but are available from the authors upon request. Please note that the parts not shown here were equal in all conditions.

On the Boundaries of Framing Terrorism: Guilt, Victimization, and the 2016 Orlando Shooting

Nathan Walter, Thomas J. Billard, and Sheila T. Murphy

The 2016 Orlando shooting offers an intriguing lens through which to evaluate the boundaries of media frames in the interpretation of terrorism. Using an experimental design ($N = 243$), the current study investigated the effects of two dominant frames—the homophobic hate crime and the Islamic terrorist frame—on collective guilt, collective victimization, and pro–lesbian, gay, bisexual, transgender, and queer (LGBTQ) political action. In addition, political partisanship and social network diversity were evaluated as potential moderators. Compared to the Islamic terrorist frame, exposure to the homophobic hate crime frame increased collective guilt and decreased collective victimization, subsequently enhancing support for the LGBTQ community. Moreover, social network diversity was shown to override the framing effect, as individuals who reported high diversity were more likely to sign a petition in solidarity with the LGBTQ community, irrespective of frame condition.

Nathan Walter (M.A., University of Haifa, 2012) is a Ph.D. candidate in the Annenberg School for Communication and Journalism at the University of Southern California. His research interests include media psychology, health communication, and political communication.

Thomas J. Billard (B.A., George Washington University, 2010) is a doctoral student in the Annenberg School for Communication and Journalism at the University of Southern California. His research interests include political communication and transgender politics.

Sheila T. Murphy (Ph.D., University of Michigan, 1990) is a Full Professor at the Annenberg School for Communication and Journalism at the University of Southern California. Her research interests include identifying the individual, interpersonal, community, ethnic, and cultural level factors that shape people's beliefs and behaviors.

In the early hours of June 12, 2016, a 29-year-old Muslim American man named Omar Mateen opened fire on the patrons of a Latin Night event at Pulse, a gay nightclub in Orlando, Florida. He killed 49 people and left 53 others wounded. This makes it, at the time of this writing, the deadliest mass shooting in U.S. history. In the days that followed the attack, journalists struggled with how to frame the event. The shooter was a Muslim man who pledged allegiance to the Islamic State in a phone call to 911, so was it an Islamic terrorist attack? The shooter was also an ostensibly heterosexual man who frequently made homophobic comments to his friends and family, so was it a homophobic hate crime? Or, as Haider (2016) asked, was it more specifically a "homophobic terror" attack? Because of the ambiguity of the event, and the consequent difficulty of attributing cause and consequence, media frames of the event became significant sources of guidance for public deliberation. Or did they?

In this study, we investigate the effects of two competing frames of the attack on Orlando's Pulse nightclub—the homophobic hate crime frame and the Islamic terrorist frame—on individuals' experiences of collective guilt and collective victimization, as well as their attitudes toward lesbian, gay, bisexual, transgender, and queer (LGBTQ) people and willingness to take pro-LGBTQ political action. This inquiry expands the literature of framing on two distinct fronts. First, in part because of the ambiguity of the circumstances of the event and the inapplicability of preexisting frames, the coverage of the Orlando attack offered competing accounts regarding the perpetrator (homophobic crime vs. Islamic terrorist) and the victim (the LGBTQ community vs. Americans). Arguably, these framing choices have direct implications for the underlying mechanism used to interpret this event. Second, the 2016 Orlando shooting brought to the fore many deeply entrenched conflicts in American culture, such as the tension between gun control and gun rights, as well as the tension between LGBTQ rights and religious conservatism, offering a unique opportunity to examine the role played by political ideology and social networks in moderating the effects of media frames.

FRAMING THE NEWS

Framing has proven to be one of the most enduring and productive paradigms in communication research. Although understandings of framing vary considerably across subfields, most perspectives converge on the idea that framing effects occur when the processing, interpretation, and retrieval of information are altered consequent the manipulation of message features (Kahneman & Tversky, 1979). Framing has been explicated as an attempt to construct social reality by providing audiences with schemas for interpreting events (Scheufele, 1999), as the selection of some elements of perceived reality and increasing their salience (Entman, 1993), and as a "discursive process of strategic actors utilizing symbolic resources to participate in collective sense–making about public issues"

(Pan & Kosicki, 2001, p. 36). Because frames define problems, diagnose causes, make moral judgments, and suggest remedies (Entman, 1993), they directly contribute to public deliberation and the formation of public opinion (Price & Tewksbury, 1997). Thus, it is not surprising that framing, together with priming and agenda-setting, has played a crucial role in the resurgence of academic interest in substantive media effects on the political process (Scheufele & Tewksbury, 2007).

This is not to suggest that journalists manipulate news content to promote particular political agendas, but rather that media frames are inevitable consequences of the attempt to convey complex and uncertain realities in an accessible, efficient, and timely manner (Gans, 1979; Scheufele & Tewksbury, 2007). Nonetheless, different framings influence perceptions of those complex and uncertain realities. Indeed, framing effects have been found to persist across diverse contexts, such as civil liberties (Nelson, Clawson, & Oxley, 1997), corporate crises (Cho & Gower, 2006), immigration (Igartua & Cheng, 2009), feminism (Terkildsen & Schnell, 1997), climate change (Wiest, Raymond, & Clawson, 2015), and terrorism (Walter, Demetriades, Kelly, & Gillig, 2016).

Given that it is virtually impossible to communicate information without offering a dominant frame, the prevalence of studies that concentrate on message design and "unique frames" has been a target of recent criticism (Borah, 2011). In particular, Cacciatore, Scheufele, and Iyengar (2016) maintained that it is time to retire the selection and salience paradigm and shift focus to equivalence framing—"a form of framing that involves manipulating the presentation of logically equivalent information" (p. 8). According to this argument, events that produce media frames with only minimal variations are particularly productive for analysis, as they potentially offer generalizable insights that go beyond the overstated assertion that exposure to content affects its interpretation (Reinhart, Marshall, Feeley, & Tutzauer, 2007). In reality, however, logically equivalent framing is rarely observed outside the laboratory, especially as, over time, interpretations tend to gravitate toward a dominant frame (Entman, 1993).

Yet, in the case of the Orlando shooting, two similar yet distinct frames competed for dominance: the homophobic hate crime frame and the Islamic terrorist attack frame (Haider, 2016). Whereas the homophobic hate crime frame identified homophobia as the cause of the attack and LGBTQ people as the victims, the Islamic terrorist attack frame identified anti-Americanism as the cause and America as the victim. Although the facts of the attack remained the same, the change in emphasis on various elements of the shooter's identities and proclaimed motives produced different frames. This was in part the case because of the "lone wolf" nature of the attack. In such contexts, when an attack is committed by a single perpetrator, though the fundamental facts tend to be clearly established and agreed on, the motives for the act remain unknown, which leaves a much wider void for interpretation (Seeger, Sellnow, & Ulmer, 2003). In the absence of a first-person rationale, questions of cause, responsibility, blame, and remedial actions are often dictated by journalists and advocacy groups to highlight a particular ideology (de Vreese, 2012). Thus, the 2016 Orlando attack provides a constructive context in which to examine framing effects by offering an observed instance of equivalence framing.

FRAMING AND COLLECTIVE EMOTIONS

Framing is rooted in the cognitive approach to social psychology (Kahneman & Tversky, 1979). However, more recent studies argue that framing effects are based on cognitive as well as *emotional* processes (Kühne, Weber, & Sommer, 2015). For example, Lecheler, Schuck, and de Vreese (2013) demonstrated that both anger and enthusiasm mediate the effects of framing on opinion toward economic policy regarding Eastern European EU members. In the context of drunk driving, Nabi (2003) found that participants exposed to an "anger frame" were twice as likely to exhibit an individual responsibility attribution, compared to those exposed to a "fear frame." Thus, whether as moderators (Kim & Niederdeppe, 2014) or mediators (Lecheler et al., 2013; Walter et al., 2016), emotions can enhance or attenuate the power of frames.

Of interest, studies have demonstrated that the effects of media frames are not restricted to individual-level emotions (e.g., fear, joy, guilt, and disgust), but frames can also vicariously arouse collective-level emotions. In particular, collective guilt and collective victimization emerge as important outcomes of framing that are relevant for intergroup relations due to their capacity to induce either self-examination or scrutiny of others (Roberts, Strayer, & Denham, 2014). This is one of the subtler, yet highly consequential, outcomes of framing, as it suggests that by highlighting certain aspects of reality, media can prompt not only ephemeral mental states (e.g., feeling anger after watching a news report or feeling sad after reading an article) but also long-lasting emotions that reshape intergroup relations (Adarves–Yorno, Jetten, Postmes, & Haslam, 2013).

Collective guilt differs from individual-level guilt in that it can be experienced even when the individual is not directly implicated in the transgression (van Leeuwen, van Dijk, & Kaynak, 2013). Individuals experience collective guilt when they realize that their ingroup has transgressed against outgroup members. As Doosje and his colleagues (1998) observed, "People can experience feelings of guilt on behalf of their group when the behavior of other ingroup members is inconsistent with norms or values of the group" (p. 873). More important, however, just as the individual-level guilty impulse is to make reparations for the caused harm (Lazarus, 1991), the collective-level guilty impulse is to make amends or reconcile with outgroup members. For example, Karaçanta and Fitness (2006) demonstrated that collective guilt can impact both attitudes and behaviors. In their study, heterosexual participants who watched a video-recorded interview with a gay student who described being physically assaulted because of his sexuality experienced collective guilt, which translated into a willingness to volunteer for a gay and lesbian antiviolence program (Karaçanta & Fitness, 2006). Similarly, Harvey and Oswald (2000) exposed White Americans to a videotape of Black civil rights protestors being abused by police and found increased support for programs that compensate Black Americans. The ability of collective guilt to entice compassion for outgroup members was also supported in studies that analyzed intergroup relationship in the context of indigenous Australians and White Australians (Halloran, 2007), American citizens and immigrants (Walter et al., 2016), and Jewish Canadians and Palestinians (Wohl & Branscombe, 2008). When such transgressions are made salient through media framing, ingroup members can experience collective guilt even though they were neither personally involved in nor responsible for the harming (Branscombe, Slugoski, & Kappen, 2004).

Although harm perpetrated by the ingroup can elicit feelings of collective guilt, harm perpetrated by outgroups to one's ingroup can elicit feelings of collective victimization. If collective guilt entails a concern for exonerating the ingroup and maintaining a positive group identity, collective victimization motivates a need for justice and is associated with actions aimed at punishing the wrongdoing outgroup (Rothschild, Landau, Molina, Branscombe, & Sullivan, 2013). In contrast to collective guilt, feelings of victimization alleviate moral concerns and serve as justifications for future transgressions. These feelings can transverse generations and result in negative emotional, attitudinal, and behavioral responses to contemporary members of the perpetrator group (Wohl & Branscombe, 2005). For example, among Jewish North Americans and indigenous Canadians, increasing the salience of participants' religious and ethnic identity resulted in more negative responses toward Germans and White Canadians, respectively (Wohl & Branscombe, 2005).

Moreover, the effects of collective victimization seem to be highly robust, such that it alleviates moral concerns regarding current transgressions, even if the events are only distantly related to the original injustice. This point was illustrated by a study that primed Canadian Jews with a memory of the Holocaust, which led to more positive perceptions of Israel's occupation of Palestinians (Wohl & Branscombe, 2008). Similarly, Americans who were reminded of the attacks of either September 11 (15 of the 19 hijackers were Saudis) or Pearl Harbor (attacked by Japanese fighter planes) experienced less empathy to the harm inflicted on the Iraqi people during the war in Iraq (Wohl & Branscombe, 2008). In the context of the terrorist attack on the French satirical magazine *Charlie Hebdo*, Walter and his colleagues (2016) showed that framing the attack as the "French September 11" led to higher levels of collective victimization among Americans, increasing support for anti-immigration policy in the United States.

Altogether, this line of research suggests that collective guilt and collective victimization are powerful catalysts that reshape attitudes, beliefs, and behaviors toward outgroup members, which occur irrespective of whether or not the person was directly involved in the transgression (O'Keefe, 2000; Schmitt, Miller, Branscombe, & Brehm, 2010). Consistent with the literature, we expect that the framing of the Orlando nightclub shooting will significantly impact whether participants experience collective guilt or collective victimization. Whereas framing the shooting as a hate crime against the LGBTQ community may make salient the historical transgressions of heterosexuals and in turn induce collective guilt, framing the shooting as a terrorist attack against the United States may make salient the ongoing conflict between the United States and so-called Islamic terrorists and, in turn, induce collective victimization.

Thus, based on the findings of past literature, we hypothesize the following:

H1a: Compared to participants in the terrorist attack frame, exposure to the hate crime frame will result in higher level of collective guilt.

H1b: Compared to participants in the terrorist attack frame, exposure to the hate crime frame will result in lower levels of collective victimization.

H1c: Compared to participants in the terrorist attack frame, exposure to the hate crime frame will result in more positive attitudes toward LGBTQ individuals.

H1d: Compared to participants in the terrorist attack frame, exposure to the hate crime frame will result in greater support for policy that would benefit LGBTQ individuals.

H1e: Compared to participants in the terrorist attack frame, exposure to the hate crime frame will increase the likelihood of signing a petition in solidarity with the LGBTQ community.

H2a: The effect of the frame condition on support for policy that would benefit LGBTQ individuals will be mediated by collective guilt.

H2b: The effect of the frame condition on support for policy that would benefit LGBTQ individuals will be mediated by collective victimization.

H2c: The effect of the frame condition on support for policy that would benefit LGBTQ individuals will be mediated by attitudes toward LGBTQ individuals.

THE LIMITATIONS OF FRAMING TERRORISM

Analyzing the limits of framing is critical to understanding the underlying mechanism that links media frames and subsequent decision making. Although increased salience can increase attention to particular issues, individuals rarely change their behavior as a result of framing, especially when dealing with highly politicized events (Hong, 2014; Niederdeppe, Shapiro, & Porticella, 2011). For instance, Bechtel and his colleagues (2015) found that, regardless of the frame with which they were presented, Swiss voters responded by increasing their support for the position of the political party with which they already identified. Similar limitations were also observed in the context of health care (Kim & Niederdeppe, 2014), support for the European Union (de Vreese, Boomgaarden, & Semetko, 2011), and gay rights (Brewer, 2003). As one might suspect, the moderating role played by preexisting beliefs is even stronger for political partisans. Thus, irrespective of the frame being used, partisans tend to act upon their extant belief systems (Lecheler & de Vreese, 2012), developed schemas (Holton, Lee, & Coleman, 2014), and party affiliations (Brewer, 2003; Hicks & Lee, 2006). In other words, compared to political moderates, partisans (on both sides) are expected to be less affected by the contextual frame and interpret the information as being consistent with their preexisting views. In the case of the Orlando shooting, this means that compared to political moderates, liberal partisans will tend to interpret the event as a hate crime, whereas conservative partisans will tend to understand the event as a terrorist attack.

Further, in the context of the Orlando nightclub shooting—an event that was not only a fatal mass shooting but also the deadliest incident of violence against LGBTQ people in U.S. history—individual understandings of the attack will likely be contingent not only on the level of political partisanship but also on social networks. Simply put, we expect that individuals who have LGBTQ people in their immediate social network will be less affected by the media frame (Herek & Capitanio, 1996). Conversely, people who are socially distant from LGBTQ individuals will be more susceptible to adopt the interpretation provided by the news coverage. Therefore, we hypothesize the following:

H3a: The framing effect will be moderated by participants' level of partisanship, such that the effect will be less pronounced for political partisans compared to their moderate counterparts.

H3b: The framing effect will be moderated by participants' social network diversity, such that the effect will be less pronounced for those with LGBTQ individuals in their immediate social networks.

METHOD

Participants and Procedure

Data for this study were collected in the United Stated on August 10, 2016, 2 months after the shooting at Pulse nightclub in Orlando, Florida. Participants were recruited through Qualtrics Pools, and they received financial compensation for their time. All participants were screened for age (older than 18), citizenship (United States citizens), English fluency, and sexual identity.[1] In total, 258 cisgender heterosexual individuals completed the questionnaire. After removing any participants with more than 15% missing data and cases that, based on the time elapsed, did not read the stimulus, data from 243 respondents were analyzed. All participants consented to take part in a study that focused on "news coverage." Then the sample was randomly assigned to either a terrorist attack frame or a hate crime frame of the same news article. Participants were subsequently presented with a questionnaire designed to measure all relevant constructs, as well as sociodemographic variables.

Material

Based on a procedure advocated in previous studies (Valkenburg, Semetko, & de Vreese, 1999), whereas both articles had an identical core, their title, opening paragraph, and closing paragraph were slightly adjusted to reflect a specific frame (terrorist attack/hate crime). To ensure that both versions of the article provided the same facts and make the information equally salient, a pilot study ($N = 26$) asked respondents to list all the substantive information about the attack, as it appeared in the two equivalent versions of the article. The stimuli were based on articles from the *Los Angeles Times* and the *New Yorker* from the weeks of the attack that provided a factual description of the event, background information about the perpetrator, and an attributed quote at the end. To reduce potential confounds, the article was presented in a purposefully vague manner, stating, "The following appeared in a newspaper that covered the June 12, 2016, Orlando shooting."

Participants in the terrorist attack condition were exposed to an article titled "An Act of Terror: The Aftermath of America's Worst Mass Shooting" (463 words), whereas the headline in the hate crime condition read "An Act of Hate: The Aftermath of America's Worst Mass Shooting" (464 words). Both versions included five paragraphs that opened with a general description of the gunman and the scene of the attack. However, a notable difference between the stimuli was the concluding paragraph, which referenced a professor of queer history in the hate crime frame and a spokesman for the Islamic State in the terrorist attack frame.

Measures

All items were measured on a 7-point Likert scale unless specified otherwise. *Attitudes toward LGBTQ individuals* were adapted from Walch, Ngamake, Francisco, Stitt, and Shingler (2012) and measured by participants' agreement

---

[1] Given potential confounds associated with varying levels of issue involvement, we decided to screen out nonheterosexual subjects ($n = 3$).

ratings with 11 statements, including "I see the LGBTQ movement as a positive thing" and "I would not mind having an LGBTQ friend" (α = .95). *Support for LGBTQ policy* was assessed by participants' agreement ratings with five proposed policies to address LGBTQ-related discrimination in the United States, such as "ending bans on LGBTQ adoption" and "overturning antisodomy laws in all states" (α = .86). *Willingness to sign a petition* was measured by prompting participants with the following: "Would you be willing to sign a petition to stand in solidarity with the Orlando LGBTQ community—and against all forms of violence, discrimination, and hate? If you choose to sign the petition, you will be redirected to the petition's website." Those who chose to sign the petition were redirected to a petition website (www.thepetitionsite.com), where they were asked to submit their full name, e-mail, and street address; participants who did not sign the petition were redirected to the end of the questionnaire. *Collective guilt* was measured with four items that composed a validated scale (Branscombe et al., 2004). The specific items included "I feel regret for our harmful past actions toward the LGBTQ community" and "I believe we should try to repair the damage that we caused to the LGBTQ community" (α = .90). *Collective victimization* was assessed with six items adopted from Branscombe, Slugoski, and Kappen's (2004) scale, which included statements such as "It upsets me that Americans suffer today because of hatred from other groups" and "It upsets me that the American way of life has been threatened by other groups through history" (α = .82).

Social network diversity was measured in two steps. Adapting the network diversity measurement employed by Hampton (2011), participants were first asked to estimate with how many people they discuss current issues (e.g., politics, health, culture, religion, business) on a regular basis. The answer options ranged from 0 to 10. In the following step, participants were asked to identify the sexual/gender identity that describes each person they listed. The answer options were "straight," "lesbian," "gay," "bisexual," "transgender," "queer," and "other." *Partisanship* was gauged with a political ideology scale, ranging from 1 (*conservative*) to 7 (*liberal*). After ensuring that the scale was normally distributed, individual scores were standardized and recoded to indicate contrasts between high and low levels of partisanship. Specifically, the first ($X < -.72$) and the fourth ($X > .90$) quartiles represented high partisanship, whereas scores within the interquartile range represented low partisanship (or moderates). The final part of the questionnaire measured participants' familiarity with the Orlando nightclub shooting, religious affiliation, education, gender, race, and income.

All analyses were conducted with SPSS v.24. Specifically, the direct effect hypotheses were assessed with independent samples t tests, ordinary least squares regressions, and a hierarchical binary logistic regression. In addition, the serial mediation/moderation hypotheses were examined using Hayes's (2013) PROCESS macro (Model 6/1; 10,000 bootstrapped samples, 95% confidence interval [CI]) and subsequently probed following the Johnson–Neyman procedure.

RESULTS

The first set of analyses looked at the descriptive characteristics of the sample across the experimental conditions. As indicated in Table 1, results show that the average participant was 40 years old, White, Christian, finished approximately 13 years of schooling, had high level of familiarity with the Orlando nightclub shooting, was not affiliated with a particular political ideology, and reported fewer than one LGBTQ individual within their immediate social network.

We next examined the main effect of framing condition on research outcomes (see Table 2). Results show a statistically significant difference across conditions for collective victimization ($d = 0.30$) and a statistically borderline effect of framing condition on collective guilt ($d = 0.22$). Of interest, no significant differences between the hate crime frame and the terrorist attack frame were found for attitudes toward LGBTQ individuals, support for pro-LGTBQ policy, or willingness to sign a petition.

To further explore how the research variables are related, zero-order correlations were computed. Table 3 displays the results. As the table shows, there were high and positive correlations between collective guilt and attitudes toward LGBTQ individuals ($r = .51$, $p < .001$) and support for pro-LGBTQ policy ($r = .48$, $p < .001$), respectively. As expected, there was a negative correlation

TABLE 1
Means, Standard Deviations, and Percentages for Research Variables

	Condition	
Variables	*Hate Crime*	*Terror Attack*
Age	41.29 (15.01)	39.90 (14.88)
Gender		
Female	51.6%	48.4%
Religion		
Christian	70.5%	75.3%
Unaffiliated	9.8%	7.4%
Atheist/agnostic	8.2%	10.7%
Muslim	0.8%	0.8%
Race		
White	79.4%	72.6%
Black	7.9%	13.7%
Hispanic	5.6%	5.6%
Familiarity with event	5.78 (1.31)	5.68 (1.44)
Political ideology	4.38 (1.86)	4.28 (1.84)
Network diversity	0.83 (1.68)	0.87 (1.59)
Education	12.98 (4.68)	12.69 (5.05)

Note. N = 243.

TABLE 2
Summary of Principal Outcome Measures by Framing Condition

Variable	Hate Crime	Terror Attack	$t(241)/\chi^2$	d/Φ_r
Collective guilt	4.70 (2.18)	4.26 (1.89)	1.65[†]	0.22
Collective victimization	3.30 (1.17)	3.66 (1.23)	2.32*	0.30
Attitudes toward LGBTQ	5.71 (1.61)	5.39 (1.70)	1.48	0.19
Support for LGBTQ policy	5.48 (1.63)	5.21 (1.69)	1.31	0.16
Sign petition: Yes	54.4%	49.6%	0.45	.004

Note. The third column provides a statistical test for the comparison between the framing conditions, using independent samples t tests (for continuous outcomes) and chi-square (for categorical outcomes). The fourth column summarizes the effect sizes. LGBTQ = lesbian, gay, bisexual, transgender, and queer.
[†]$p < .10$. *$p < .05$.

TABLE 3
Means, Standard Deviations, and Zero-Order Correlations

Variable	M	SD	1	2	3	4	5
1. Collective guilt	4.49	2.05	—				
2. Collective victimization	3.48	1.22	−.32***	—			
3. Attitudes	5.56	1.66	.51***	−.19**	—		
4. Support for policy	5.35	1.59	.48***	−.17**	.73***	—	
5. Network diversity	0.85	1.62	.11[†]	−.11[†]	.17**	.05	—

Note. N = 243.
[†]$p < .10$. **$p < .01$. ***$p < .001$.

between collective guilt and collective victimization ($r = -.32$, $p < .001$). Likewise, concurring with previous research (Walter et al., 2016), collective victimization was negatively associated with attitudes ($r = -.19$, $p < .01$) and support for policy ($r = -.17$, $p < .01$), respectively. In addition, the correlation between network diversity and collective guilt ($r = .11$, $p < .10$) and collective victimization ($r = -.11$, $p < .10$), respectively, were nonsignificant. Finally, a weak, albeit significant correlation was estimated between network diversity and attitudes toward LGBTQ individuals ($r = .17$, $p < .01$).

H1 and H2 were tested using PROCESS (Model 6 set at 10,000 bootstrapped samples with CI of 95%), an ordinary least squares regression that provides unstandardized estimates (for direct effects, see Figure 1). In total, the path model results offer support for the hypotheses. Specifically, as expected, exposure to the Orlando nightclub shooting through the hate crime frame increased collective guilt ($\beta = .59$, $SE = .26$), which enhanced supportive attitudes toward LGBTQ individuals ($\beta = .41$, $SE = .05$), resulting in higher levels of support for

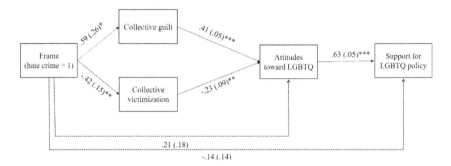

FIGURE 1 Unstandardized coefficients (and standard errors in parentheses) for the direct effects of framing condition on collective guilt; collective victimization; attitudes toward lesbian, gay, bisexual, transgender, and queer individuals (LGBTQ); and support for LGBTQ policy. *Note.* *$p < .05$. **$p < .01$. ***$p < .001$.

pro-LGBTQ policy ($\beta = .63$, $SE = .05$). Likewise, the hate crime frame decreased collective victimization ($\beta = -.42$, $SE = .15$), which was in turn a negative predictor of favorable attitudes toward LGBTQ individuals ($\beta = -.23$, $SE = .09$).

Furthermore, H1e predicted that exposure to the hate crime frame would increase the likelihood of participants agreeing to sign a petition of solidarity with the LGBTQ community. This hypothesis was assessed with a hierarchical binary logistic regression. Framing condition was entered at Block 1, collective guilt and collective victimization were added in Block 2, attitudes and support for policy were introduced in Block 3, and the role of social network diversity was estimated in Block 4. Contrary to our expectation, controlling for other variables in the model, exposure to the hate frame condition was not a significant predictor of signing the petition (odds ratio $[OR] = 1.10$, $p = .75$), 95% CI [.62, 1.95].[2] Of interest, collective victimization ($OR = 0.68$, $p = .005$), 95% CI [.44, .89], and attitudes toward LGBTQ individuals ($OR = 1.45$, $p = .01$), 95% CI [1.09, 1.94], were significant predictors such that, on average, collective victimization decreased the likelihood of signing the petition, whereas attitudes toward LGBTQ individuals increased the likelihood of signing the petition. Yet the best predictor of the likelihood of agreeing to sign the petition was the composition of participants' social networks ($OR = 2.88$, $p = .005$), 95% CI [1.74, 4.77]. Simply put, on average, having one additional LGBTQ person in one's immediate social network increased the likelihood of signing the petition

---

[2] $OR = 1$ indicates that the predictor does not affect the odds of the outcome. $OR > 1$ indicates that the predictor is associated with higher odds of the outcome. $OR < 1$ indicates that the predictor is associated with lower odds of the outcome.

by nearly 3 times. See Table 4 for a full layout of the logistic regression, including odds ratios and 95% CIs.

To examine H2, we used PROCESS (Model 6 set at 10,000 bootstrapped samples with CI of 95%). In agreement with our hypotheses, the effect of the frame condition on support for pro-LGBTQ policy was significantly mediated through collective guilt and attitudes toward LGBTQ individuals ($b = .15$, $SE = .07$), 95% CI [.03, .31], as well as through collective victimization and attitudes toward LGBTQ individuals ($b = .06$, $SE = .04$), 95% CI [.01, .18].

Finally, using PROCESS (Model 1 set at 10,000 bootstrapped samples), we examined whether the relationship between frame condition and collective guilt/ collective victimization varied by social network diversity and partisanship. First, results showed a statistically significant Frame Condition × Social Network interaction on collective guilt ($b = -.19$, $SE = .11$, $p < .05$). The analysis also found a statistically significant Frame Condition × Social Network interaction on collective victimization ($b = .13$, $SE = .07$, $p < .05$). As Figure 2 indicates, an increase in the number of LGBTQ people within participants' social networks tipped the effect both for collective guilt and for collective victimization. More important, probing the interaction with the Johnson–Neyman technique indicated that those who listed fewer than 1.15 LGBTQ individuals within their social

TABLE 4

95% Confidence Interval (CI) for Odds Ratio for the Prediction of Signing a LGBTQ Solidarity Petition by Research Variables

Predictor	Block 1 OR	Block 2 OR	Block 3 OR	Block 4 OR
Intercept	.86	.08	.01	.01
Hate-crime frame	1.05	1.01	1.10	1.10
	[.61, 1.84]	[.58, 1.76]	[.62, 1.95]	[.62, 1.95]
Collective guilt		1.29***	1.15†	1.15
		[1.11,1.49]	[.97, 1.37]	[.97,1.37]
Collective victimization		.71**	.69**	.68**
		[.54, .96]	[.54, .93]	[.54, .91]
Attitudes toward LGBTQ			1.51***	1.45**
			[1.13, 2.01]	[1.09, 1.94]
Support for policy			.96	.97
			[.74, 1.25]	[.75, 1.29]
Network diversity				2.88***
				[1.74, 4.77]
R^2 (Cox & Snell/Nagelkerke)	.001/.001	.07/.09	.12/.16	.22/.29
χ^2 (Hosmer & Lemeshow)	.001	3.77	4.38	8.49
Model χ^2	.001	15.27***	28.34***	58.83***

Note. Odds ratios (ORs) are shown for binary outcomes: 95% CI for ORs are in brackets. LGBTQ = lesbian, gay, bisexual, transgender, and queer.

$^\dagger p < .10$. **$p < .01$. ***$p < .001$.

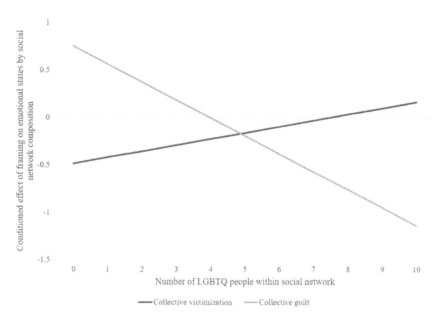

FIGURE 2 The conditioned effect of framing on collective guilt and collective victimization by social network diversity. *Note.* LGBTQ = lesbian, gay, bisexual, transgender, and queer.

network (77.19% of the sample) showed a significant positive effect of the frame condition on collective guilt, whereas the effect was nonsignificant for the remaining 22.81% sample. Likewise, 1.92 (or fewer) LGBTQ individuals within participants' social networks (72.64% of the sample) was the cutoff score for a significant effect of frame condition on collective victimization. With respect to the moderation effect of partisanship, the analysis revealed nonsignificant interactions both for collective guilt ($b = -.51$, $SE = .53$, $p = .34$) and for collective victimization ($b = .47$, $SE = .41$, $p = .31$).

DISCUSSION

The current study used coverage of the Orlando nightclub shooting to demonstrate how media frames (homophobic hate crime and Islamic terrorist attack) cement competing interpretations by evoking social categories and collective emotions. Specifically, increasing ingroup inclusiveness—from an attack on the LGBTQ community to an attack on Americans—leads to greater collective victimization, subsequently undermining the need for reconciliation with the LGBTQ outgroup. Among participants exposed to the homophobic hate crime frame, we observed an opposite pattern of results, whereby framing the victim as the LGBTQ community increased collective guilt, encouraging participants to make reparations with the victims. It is important to note that these responses are not incidental, as they are motivated in service of reducing the social identity threat associated with exposure to either the misdeeds of the ingroup or the acceptance of harmful actions toward the ingroup (Rotella & Richeson, 2013).

The first and second hypotheses in the study assumed that, compared to the terrorist attack frame, exposure to a homophobic hate crime framing of the event will induce greater collective guilt, less collective victimization, resulting in more favorable attitudes toward the LGBTQ community and higher support toward policies that would benefit LGBTQ individuals. Even though the main effect analysis did not record any significant effects for the attitudinal outcomes, the path analysis supported our hypotheses. More specifically, exposure to the Orlando attack through the homophobic hate crime frame predicted support for the LGBTQ community, as participants tended to see their ingroup not as a victim but rather as the perpetrator against LGBTQ individuals.

Of interest, as often happens in social research, attitudes did not predict behavior—neither frame increased the likelihood of participants signing a petition in solidarity with the LGBTQ community. There are several factors that can perhaps explain people's hesitation to sign the petition. First, irrespective of the frame condition, people may think that their signature will not matter. Although political internal efficacy was not directly assessed in the present study, recent national reports suggest that, overall, 61% of the public falls into the low to medium political efficacy categories, with less educated White men being the least efficacious group (Pew Research Center, 2015). Keeping in mind that participants in the current study were predominately White and that nearly half of them were male, it stands to reason that low political efficacy played a role in people's decision not to sign. Second, given that the questionnaire was administered online and that in order to sign the petition respondents were redirected to a new website, perhaps people declined to sign the petition because they were concerned with being able to finish the survey. Relatedly, it can be argued that participants simply opted to take the shorter route to finishing the study, knowing that they will not receive additional compensation for signing a petition. Finally, another methodological explanation is associated with a potential ceiling/floor effect, namely, willingness to sign the petition was measured with two binary answer options (yes/no), thus more variance in this measurement would possibly have produced a significant effect.

The inconsistencies between the attitudinal outcomes and the behavioral outcomes shift the focus away from the direct effects of framing toward trying to analyze individual differences that can facilitate the role played by media frames. The third hypothesis attempted to test the limitations of the framing effect by assessing political partisanship and social network diversity as potential moderators. Based on previous research, we expected political partisans to rely on preexisting beliefs and established schemas rather than the contextual frame (Powlick & Katz, 1998). In reality, we found no support for this assertion, suggesting that the frame conditions tended to overpower existing belief systems. The failure to support this hypothesis may be due to what Haider (2016) recognized as the tension that arises when events do not necessary fit within a common framework. That is, although there are separate frames for homophobic hate crimes and Islamic terrorism, *homophobic terrorism* is much more difficult to pin down. Thus, preestablished schemas were of limited use when trying to categorize an unprecedented event.

Conversely, social network diversity was a significant moderator both for collective guilt and for collective victimization, suggesting that having LGBTQ individuals in one's immediate social network diminished the framing effect, putting more emphasis on people's interpersonal relationships rather than contextual frames. Using the binary logistic regression, the findings also indicate that social network diversity was the strongest predictor of signing the petition, as participants were 3 times more likely to comply for each additional LGBTQ individual within their social network. This is an area that deserves more theoretical and empirical attention. On one hand, in line with traditional approaches to media effects (Katz & Lazarsfeld, 1955), the prioritization of personal influence compared to mass media highlights considerable limitations of framing. Namely, if ideas that flow from the mass media are mediated by interpersonal discussion, contextual media frames seem to lose part of their potency. On the other hand, given that homophily is a firmly established organizing mechanism of social networks, interpersonal relationships with outgroup members are often not a feasible option. For example, in the current study, on average, participants had fewer than one LGBTQ individual in their social network. In that light, media frames are highly influential, especially when covering events involving outgroups with whom ingroup members rarely interact.

There are several limitations that should be carefully acknowledged. First, the study was conducted 2 months after the Orlando nightclub shooting, and it offers only a single snapshot of a much longer time line of news coverage and public discourse. Thus, based on these measurements, it is impossible to determine how the framing effect unfolds over time, as more information becomes available to the public. For instance, one can speculate that in the immediate aftermath of terrorist attacks, media frames play a larger role in communicating the social reality, yet over time, interpersonal relations and secondary sources overpower the initial interpretations provided by media sources. Although this argument would be consistent with the media system dependency approach (Ball–Rokeach, 1985), it remains to be empirically evaluated. In addition, the current study did not employ a manipulation check, which could have helped ensure that the terrorist attack and the hate crime frames were actually interpreted as such.

Another limitation is related to the fact that our study focused only on a single event, which limits our ability to generalize these results to the coverage of other events that involve the use of competing frames. Indeed, future studies could apply the proposed theoretical mechanisms to other contexts, further examining the interplay between media frames, collective level emotions, and social identities. Another consideration related to external validity is the fact that we did not account for selective exposure. Simply put, one of the main assumptions of equivalence framing is that individuals have an equal chance to be exposed to either of the frames. Yet the documented need for opinion reinforcement may shape individuals' preferences for particular frames and the avoidance of frames that challenge their opinions (Garrett, 2009).

An additional limitation has to do with the question of causality. The interpretation of results offered in this article suggests that exposure to coverage regarding the Orlando nightclub shooting elicited an emotional response, subsequently affecting participants' attitudes toward the LGBTQ community. However, an equally plausible interpretation would maintain that framing directly affected attitudinal outcomes, which ultimately exerted an influence on collective guilt and collective victimization. Given that the mediators are not manipulated but rather simply measured, it is impossible to account for various alternative explanations (Pearl, 2014). Last, it is important to acknowledge that the present study did not utilize a representative sample, and thus it is hard to estimate the external validity of the results. Although the distribution for most variables seems to correspond with the general American population, the generalizability of other variables is harder to estimate. For instance, it is unclear whether the average American talks about current affairs with more or less nonheterosexual individuals than the average participant in the current study, who reported on talking with fewer than one LGBTQ individual. To some extent, if treating the prevalence of U.S. adults who identify as LGBT as a proxy for our measurement, then the distribution in the sample is relatively representative (according to Gates & Newport, 2012, 3.4% Americans identify themselves as lesbian, gay, bisexual, or transgender). With that in mind, it is hard to speculate the true distribution of discussions with LGBTQ individuals.

In closing, framing emerges as an important resource not only at the immediate aftermath of terrorist attacks but also in the sense-making processes that follow. Yet, excepting notable examples (Lecheler & de Vreese, 2012), the literature on framing has been largely silent about the interplay between media frames and other information resources available in individuals' communication ecologies. As demonstrated, these resources can be key elements in the acceptance of news coverage and its subsequent effect on political behavior. Moreover, the current study echoes the call made by Cacciatore et al. (2016) to reinvigorate framing research by proposing a more holistic approach to the study of framing and terrorism, one that is anchored in media effects, collective memory, and social networks.

REFERENCES

Adarves-Yorno, I., Jetten, J., Postmes, T., & Haslam, S. A. (2013). What are we fighting for? The effects of framing on ingroup identification and allegiance. *Journal of Social Psychology*, *153*(1), 25–37. doi:10.1080/00224545.2012.701673

Ball-Rokeach, S. J. (1985). The origins of individual media–system dependency: A sociological framework. *Communication Research*, *12*(4), 485–510. doi:10.1177/009365085012004003

Bechtel, M., Hainmueller, J., Hangartner, D., & Helbling, M. (2015). Reality bites: The limits of framing effects for salient and contested policy issues. *Political Science Research and Methods*, *3* (3), 683–695. doi:10.2139/ssrn.2025552

Borah, P. (2011). Conceptual issues in framing theory: A systematic examination of a decade's literature. *Journal of Communication*, *61*(2), 246–263. doi:10.1111/j.1460–2466.2011.01539.x

Branscombe, N. R., Slugoski, B., & Kappen, D. M. (2004). *Collective guilt: What it is and what it is not*. Cambridge, UK: Cambridge University Press.

Brewer, P. R. (2003). The shifting foundations of public opinion about gay rights. *Journal of Politics*, *65*(4), 1208–1220.

Cacciatore, M. A., Scheufele, D. A., & Iyengar, S. (2016). The end of framing as we know it … and the future of media effects. *Mass Communication and Society*, *19*(1), 7–23. doi:10.1080/ 15205436.2015.1068811

Cho, S. H., & Gower, K. K. (2006). Framing effect on the public's response to crisis: Human interest frame and crisis type influencing responsibility and blame. *Public Relations Review*, *32*(4), 420–422. doi:10.1016/j.pubrev.2006.09.011

de Vreese, C. H. (2012). New avenues for framing research. *American Behavioral Scientist*, *56*(3), 365–375. doi:10.1177/0002764211426331

de Vreese, C. H., Boomgaarden, H. G., & Semetko, H. A. (2011). (In)direct framing effects: The effects of news media framing on public support for Turkish membership in the European Union. *Communication Research*, *38*(2), 179–205. doi:10.1177/0093650210384934

Doosje, B., Branscombe, N. R., Spears, R., & Manstead, A. S. R. (1998). Guilty by association: When one's group has a negative history. *Journal of Personality and Social Psychology*, *75*(4), 872–886. doi:10.1037/0022-3514.75.4.872

Entman, R. M. (1993). Framing: Toward clarification of a fractured paradigm. *Journal of Communication*, *43*(4), 51–58. doi:10.1111/j.1460–2466.1993.tb01304.x

Gans, H. J. (1979). *Deciding what's news: A study of CBS evening news, NBC nightly news, newsweek, and time*. Evanston, IL: Northwestern University Press.

Garrett, R. K. (2009). Echo chambers online? Politically motivated selective exposure among Internet news users. *Journal of Computer–Mediated Communication*, *14*(2), 265–285. doi:10.1111/j.1083–6101.2009.01440.x

Gates, G. J., & Newport, F. (2012, October 18). *Special report: 3.4% of U.S. adults identify as LGBT*. Retrieved from http://www.gallup.com/poll/158066/special-report-adults-identify-lgbt.aspx

Haider, S. (2016). The shooting in Orlando, terrorism or toxic masculinity (or both?). *Men and Masculinities*, *19*(5), 555–565. doi:10.1177/1097184X16664952

Halloran, M. J. (2007). Indigenous reconciliation in Australia: Do values, identity and collective guilt matter? *Journal of Community & Applied Social Psychology*, *17*, 1–18.

Hampton, K. N. (2011). Comparing bonding and bridging ties for democratic engagement. *Information, Communication & Society*, *14*(4), 510–528.

Harvey, R. D., & Oswald, D. L. (2000). Collective guilt and shame as motivation for White support of Black programs. *Journal of Applied Social Psychology*, *30*(9), 1790–1811.

Hayes, A. F. (2013). *Introduction to mediation, moderation, and conditional process analysis: A regression–based approach*. New York, NY: Guilford Press.

Herek, G. M., & Capitanio, J. P. (1996). "Some of my best friends": Intergroup contact, concealable stigma, and heterosexuals' attitudes toward gay men and lesbians. *Personality and Social Psychology Bulletin*, *22*(4), 412–424.

Hicks, G. R., & Lee, -T.-T. (2006). Public attitudes toward gays and lesbians. *Journal of Homosexuality*, *51*(2), 57–77.

Holton, A., Lee, N., & Coleman, R. (2014). Commenting on health: A framing analysis of user comments in response to health articles online. *Journal of Health Communication*, *19*(7), 825–837. doi:10.1080/10810730.2013.837554

Hong, T. (2014). Examining the role of exposure to incongruent messages on the effect of message framing in an Internet health search. *Communication Research*, *41*(2), 159–179. doi:10.1177/0093650212439710

Igartua, J. J., & Cheng, L. (2009). Moderating effect of group cue while processing news on immigration: Is the framing effect a heuristic process? *Journal of Communication*, *59*(4), 726–749. doi:10.1111/j.1460-2466.2009.01454.x

Kahneman, D., & Tversky, A. (1979). Prospect theory: An analysis of decision under risk. *Econometrica*, *47*(2), 263–292.

Karaçanta, A., & Fitness, J. (2006). Majority support for minority out-groups: The roles of compassion and guilt. *Journal of Applied Social Psychology*, *36*, 2730–2749.

Katz, E., & Lazarsfeld, P. F. (1955). *Personal influence: The part played by people in the flow of mass communication*. New York, NY: The Free Press.

Kim, S. J., & Niederdeppe, J. (2014). Emotional expressions in antismoking television advertisements: Consequences of anger and sadness framing on pathways to persuasion. *Journal of Health Communication*, *19*(6), 692–709. doi:10.1080/10810730.2013.837550

Kühne, R., Weber, P., & Sommer, K. (2015). Beyond cognitive framing processes: Anger mediates the effects of responsibility framing on the preference for punitive measures. *Journal of Communication*, *65*(2), 259–279. doi:10.1111/jcom.12151

Lazarus, R. S. (1991). *Emotion and adaptation*. New York, NY: Oxford University Press.

Lecheler, S., & de Vreese, C. H. (2012). News framing and public opinion: A mediation analysis of framing effects on political attitudes. *Journalism & Mass Communication Quarterly*, *89*(2), 185–204. doi:10.1177/1077699011430064

Lecheler, S., Schuck, A. R. T., & de Vreese, C. H. (2013). Dealing with feelings: Positive and negative discrete emotions as mediators of news framing effects. *Communications*, *38*(2), 189–209. doi:10.1515/commun-2013-0011

Nabi, R. L. (2003). Exploring the framing effects of emotion: Do discrete emotions differentially influence information accessibility, information seeking, and policy preference? *Communication Research*, *30*(2), 224–247. doi:10.1177/0093650202250881

Nelson, T., Clawson, R., & Oxley, Z. (1997). Media framing of a civil liberties conflict and its effect on tolerance. *American Political Science Review*, *91*(3), 567–584.

Niederdeppe, J., Shapiro, M. A., & Porticella, N. (2011). Attributions of responsibility for obesity: Narrative communication reduces reactive counterarguing among liberals. *Human Communication Research*, *37*(3), 295–323. doi:10.1111/j.1468-2958.2011.01409.x

O'Keefe, D. J. (2000). Guilt and social influence. *Communication Yearbook*, *23*, 67–101.

Pan, Z., & Kosicki, G. M. (2001). Framing as a strategic action in public deliberation. In S. D. Reese, O. H. Gandy, & A. E. Grant (Eds.), *Framing public life: Perspectives on media and our understanding of the social world* (pp. 35–66). Mahwah, NJ: Erlbaum.

Pearl, J. (2014). Interpretation and identification of causal mediation. *Psychological Methods*, *19*(4), 459–481. doi:10.1037/a0036434

Pew Research Center. (2015). *Perceptions of the public's voice in government and politics*. Retrieved from http://www.people-press.org/2015/11/23/8-perceptions-of-the-publics-voice-in-government-and-politics/

Powlick, P. J., & Katz, A. Z. (1998). Defining the American public opinion/foreign policy nexus. *Mershon International Studies Review, 42*(1), 29–61. doi:10.1111/1521–9488.00091

Price, V., & Tewksbury, D. (1997). News values and public opinion: A theoretical account of media priming. In G. Barnett, & F. J. Boster (Eds.), *Progress in the communication sciences* (Vol. 13, pp. 173–212). Greenwich, UK: Ablex.

Reinhart, A. M., Marshall, H. M., Feeley, T. H., & Tutzauer, F. (2007). The persuasive effects of message framing in organ donation: The mediating role of psychological reactance. *Communication Monographs, 74*(2), 229–255. doi:10.1080/03637750701397098

Roberts, W., Strayer, J., & Denham, S. (2014). Empathy, anger, guilt: Emotions and prosocial behaviour. *Canadian Journal of Behavioural Science, 46*(4), 465–474.

Rotella, K. N., & Richeson, J. A. (2013). Motivated to "forget": The effects of in–group wrongdoing on memory and collective guilt. *Social Psychological and Personality Science, 4*(6), 730–737. doi:10.1177/1948550613482986

Rothschild, Z. K., Landau, M. J., Molina, L. E., Branscombe, N. R., & Sullivan, D. (2013). Displacing blame over the ingroup's harming of a disadvantaged group can fuel moral outrage at a third–party scapegoat. *Journal of Experimental Social Psychology, 49*(5), 898–906. doi:10.1016/j.jesp.2013.05.005

Scheufele, D. A. (1999). Framing as a theory of media effects. *Journal of Communication, 49*(1), 103–122.

Scheufele, D. A., & Tewksbury, D. (2007). Framing, agenda setting, and priming: The evolution of three media effects models. *Journal of Communication, 57*(1), 9–20.

Schmitt, M. T., Miller, D. A., Branscombe, N. R., & Brehm, J. W. (2010). The difficulty of making reparations affects the intensity of collective guilt. *Group Processes and Intergroup Relations, 11*, 267–279. doi:10.1177/1368430208090642

Seeger, M. W., Sellnow, T. L., & Ulmer, R. R. (2003). *Communication and organizational crisis.* Westport, CT: Greenwood.

Terkildsen, N., & Schnell, F. (1997). How media frames move public opinion: An analysis of the women's movement. *Political Research Quarterly, 50*(4), 879–900.

Valkenburg, P. M., Semetko, H. A., & de Vreese, C. H. (1999). The effects of news frames on readers' thoughts and recall. *Communication Research, 26*(5), 550–569. doi:10.1177/009365099026005002

van Leeuwen, E., van Dijk, W., & Kaynak, U. (2013). Of saints and sinners: How appeals to collective pride and guilt affect outgroup helping. *Group Processes & Intergroup Relations, 16* (6), 781–796. doi:10.1177/1368430213485995

Walch, S. E., Ngamake, S. T., Francisco, J., Stitt, R. L., & Shingler, K. A. (2012). The attitudes toward transgendered individuals scale: Psychometric properties. *Archives of Sexual Behavior, 41* (5), 1283–1291. doi:10.1007/s10508–012–9995–6

Walter, N., Demetriades, S. Z., Kelly, R., & Gillig, T. K. (2016). Je suis Charlie? The framing of ingroup transgression and the attribution of responsibility for the Charlie Hebdo attack. *International Journal of Communication, 10*, 3956–3974.

Wiest, S. L., Raymond, L., & Clawson, R. A. (2015). Framing, partisan predispositions, and public opinion on climate change. *Global Environmental Change, 31*, 187–198. doi:10.1016/j.gloenvcha.2014.12.006

Wohl, M. J., & Branscombe, N. R. (2005). Forgiveness and collective guilt assignment to historical perpetrator groups depend on level of social category inclusiveness. *Journal of Personality and Social Psychology, 88*(2), 288–303. doi:10.1037/0022–3514.88.2.288

Wohl, M. J., & Branscombe, N. R. (2008). Remembering historical victimization: Collective guilt for current ingroup transgressions. *Journal of Personality and Social Psychology, 94*(6), 988–1006. doi:10.1037/0022–3514.94.6.988

Proximity and Terrorism News in Social Media: A Construal-Level Theoretical Approach to Networked Framing of Terrorism in Twitter

K. Hazel Kwon⊚, Monica Chadha, and Kirstin Pellizzaro

This study investigates networked framing of terrorism news in Twitter by distinguishing three proximity effects (geographic, social, and temporal proximity) on audience and media institutional frames (i.e., episodic/thematic and space frames), based on construal-level theory. An analysis of tweets during the Boston Marathon bombing and the Brussels Airport attack finds that institutional and audience frames show similarity but do not always converge on Twitter. Similarities in the audience and institutional frames are attributed to a universal human tendency for social categorization, inherent in the minds of not only ordinary citizens but also journalists. Proximity effects, however, were more salient on audience frames than on institutional frames.

K. Hazel Kwon (Ph.D., SUNY-Buffalo, 2011) is an assistant professor in the Walter Cronkite School of Journalism and Mass Communication at Arizona State University. Her research interests include collective sense-making in social media, anti-social behavior in cyberspace, and social network influence.

Monica Chadha (Ph.D., University of Texas at Austin, 2014) is an assistant professor in the Walter Cronkite School of Journalism and Mass Communication at Arizona State University. Her research interests include news media startups, entrepreneurial journalism and local/community news.

Kirstin Pellizzaro (M.S., Southern Illinois University at Edwardsville, 2011) is a doctoral student in the Walter Cronkite School of Journalism and Mass Communication at Arizona State University. Her research interests include social media, parasocial interaction, and newscasters' professional behaviors through illness and hardship.

When the explosion of two bombs on Boylston Street in Boston—a trailhead to the 2013 Boston Marathon's finishing line—killed three civilians and injured 264 others, social media users in the United States and other nations expressed sadness, fear, outgroup hostility, and various attributions of responsibility. Three years later in 2016, another terrorist attack at the Brussels Airport in Belgium resulted in 33 civilian deaths and more than 300 injured. People took to social media yet again to express their emotions and opinions, not only within Belgium but also in other European countries and the United States.

Social media platforms have become popular channels through which global audiences receive, share, and discuss relevant news topics including terrorism events. Although terrorism has been a news topic with high currency value for decades, media coverage on terrorism has become particularly salient in the United States since the 9/11 attack in 2001 (Norris, Kern, & Just, 2003). Moreover, several global attacks thereafter have ensured terrorism news occupies a prominent position not only in the media of the attacked country but also in other "culturally proximate" countries, reinstating the role of news coverage in shaping domestic perception of global threats (Nossek & Berkowitz, 2006, p. 694).

This study explores the global nature of public's terrorism sense-making in the contemporary social media environment. Specifically, it aims to advance terrorism framing literature in two ways. First, we revisit "proximity" as a news value in the context of social media framing of terrorism. Proximity is one of the conventional indicators of newsworthiness that influences the ways in which foreign news events are reported, including terrorist attacks (Nossek & Berkowitz, 2006). A premise has been that a distant event will be either not covered or covered only superficially because of lack of newsworthiness. In the social media milieu, however, audiences are often exposed to and willing to engage with faraway crisis events in real-time, easily transcending geographic and editorial boundaries (Kwon, Xu, Wang, & Chon, 2016). Especially, an increased awareness of international terrorism is known to have some impact on shifting domestic public opinions (Finseraas & Listhaug, 2013). Therefore, it is timely to reconsider proximity effects on terrorism news framing among networked social media publics. For this goal, we elaborate the notion of proximity along three axes—geography, time, and social closeness—by borrowing Trope and Libermann's (2010) discussion of "psychological distance" (p.440), a core concept in their construal-level theory (CLT).

Second, we distinguish the two types of social media actors who engage with networked framing—media and institutional actors, and ordinary citizens—and compare their framing on terrorism. The notion of "networked framing" has insightfully addressed the interplay between narratives of media institutions and general publics in shaping news stories in networked environments (Meraz & Papacharissi, 2013). Nonetheless, few studies have actually compared the differences or similarities between institutional frames (e.g., media organization, government institutions) and organic, spontaneously occurring audience frames in social media platforms. The comparison between these two frames are especially pertinent in the context of terrorism because terrorism news coverage is known to have direct impact on public perceptions of policies such as national security (Davis & Silver, 2004), civil liberties (Huddy, Khatib, & Capelos, 2002), and intercultural relations (Das, Bushman, Bezemer, Kerkhof, & Vermeulen, 2009). The particular focus of this study is on the ways in which proximity-as-news-value influences audience frames of terrorism, which is juxtaposed with media institutional frames.

For the empirical analyses of similarities and differences in narratives of media organizations and audiences, we examine Twitter. This platform is a popular space where ordinary citizens can and do easily express their feelings and perspectives on terrorism events. Thus, networked frames on Twitter may manifest public understanding of this issue. Two questions are addressed: First, how does proximity influence the ways in which terrorism is framed on social media platforms, specifically Twitter? Second, how does social media audience's framing differ from institutional framing of terrorism? Statistical analyses were carried out on Twitter samples collected during two recent terrorism events: the Boston Marathon bombing in 2013 (hereupon, BMB13) and Brussels Airport bombing in 2016 (BAB16). Overall findings suggest that despite some similarities between institutional and audience frames, differences exist in terms of proximity effects on frames.

BACKGROUND

Understanding Terrorism in Western Democracies

Terrorism refers to "the systematic use of coercive intimidation against civilians for political goals" (Norris et al., 2003, p. 2). Public opinion on terrorism is often closely interwoven with citizens' overall assessment of sociocultural values. For example, the World Value Survey (2015) suggests that U.S. citizens in the post-9/11 decade have increasingly stressed "survival" values with an emphasis on security and ethnocentric stances while drifting away from "self-expression" values that give priority to diversity and democratic rights. The terrorism rhetoric of "safer society" also influences political decision making, facilitating the trade-off between liberties and security (Davis & Silver, 2004; Huddy et al., 2002).

Media and institutional frames such as "war on terrorism" have contributed to the heightened sense of threat in public minds and bolstered citizen support for preemptive measures related to national defense, including unilateral foreign policy (Lewis & Reese, 2009) and domestic wiretapping (Landau, 2011). Such fear-invoking frames, however, are contradictory to official statistics that have tracked the number of terror-related incidents that took place between 1985 and 2000 and between 2001 and 2015 (National Consortium for the Study of Terrorism and Responses to Terrorism, 2015). During 1985–2000, terrorist acts in the United States and Western Europe accounted for 1.9% and 13.5% of the total attacks worldwide. From 2001 to 2015, however, the proportion dropped sharply to account for only 0.4% (the United States), and 2.75% (Western Europe) of terrorist attacks worldwide (National Consortium for the Study of Terrorism and Responses to Terrorism, 2015). This discrepancy between number of terrorism incidents and heightened public fear is partly attributable to news media effects on public construction of terrorism (Dalal, 2017).

Media Institutional Framing of Terrorism

Terrorism has a tremendous impact on the news agenda—regardless of specific framing patterns chosen by the media—triggering an "unconscious death anxiety," a primer of fear-driven judgment and prejudice (Das et al., 2009, p. 455). News framing is a process of meaning construction based on a series of "organizing principles" (Reese, 2001, p. 11). Even if the same terrorism event is reported, news can induce dissimilar public opinions depending on the level of emotional appeal in the story narrative. Audiences were increasingly supportive of the "hawkish" policy when they were exposed to high fear-inducing cues, whereas support for a "dovish" policy remained consistent when exposed to information-centric and neutral toned news coverage (Gadarian, 2010, p. 471).

In general, terrorism news research has centered on the organizing principles held by "elite" media or professional journalists (e.g., Fahmy, 2010; Iyengar, 1994; Morin, 2016; Nossek & Berkowitz, 2006). There is consensus that media frames constructed as a result of journalists' ideology, gatekeeping practices, and their emphasis on newsworthiness, determine audiences' views on terrorism (Iyengar, 1994; Lewis & Reese, 2009).

Audience Framing of Terrorism

Framing research has matured based on two methodological branches that examine relationships between news frames and audience responses. One is the "idealistic approach" based primarily on experiments, and the other is the "pragmatic approach," based on survey data or public opinion polls (McLeod & Shah, 2015, p. 15). The idealist approach ensures internal validity, whereas the pragmatic approach is better suited for research that emphasizes ecological validity. Nonetheless, both approaches lack perspective in naturally occurring, spontaneous discourse among audiences on social media platforms. Therefore, examining audience framing of terrorism may further elucidate whether public frames converge with media institutional frames.

Audience frames are especially pertinent to terrorism research because some studies have suggested it is *not* always the case that media directly represents public assessment of an issue. For example, Nacos and Torres-Reyna (2003) found that news coverage of the Muslim population after 9/11 had become increasingly positive and included more diverse thematic issues (e.g., civil liberties) than the pre-9/11 period. However, a public poll in post-9/11 conversely indicated increased biases of U.S. citizens toward American Muslims. Another nationwide survey post-9/11 found although news coverage emphasized aspects of national security and terrorist individuals or groups in their news frames, these features were relatively infrequent in the public responses (Traugott & Brader, 2003). Instead, the authors found that hatred-based attribution of responsibility and foreign policy issues were dominant in the public frames.

In addition, social media has further evolved the concept of news framing. The audience is no longer a silent actor in what Meraz and Papacharissi (2013) called "networked framing," in which the act of news redistribution by online users not only sets frames but also builds or evolves existing frames (p. 6). Audiences on Twitter can retweet an elite media frame or, in most cases, add their own narrative to the current frame, thus evolving or creating a new frame. Meraz and Papacharissi found in a discourse analysis of the 2011 Egyptian uprising, frames by prominent actors on Twitter "were persistently revised, rearticulated, and re-dispersed by both crowd and elite" (p. 138). In addition, a study of China's social media site Weibo found evidence of networked framing; retweets from audiences often revised the frame of the original post creating new definitions or diagnosis of a given event (Nip & Fu, 2016). This evolution of frames points to the importance of understanding the formation of public opinion on social media.

Also, networked framing points to the interaction of media or institutional actors with the general public on social media. For instance, popular hashtags often emerge from the public in a bottom-up manner, but popularity is gained through both elite and nonelite use of these hashtags (González-Ibánez, Muresan,

& Wacholder, 2011). Hashtags can produce dominant, thematic frames that shape the narrative of an event while it unfolds. Furthermore, with increased and extensive use, they can "thus enact, enable, and sustain the framing of select interpretations, aspects, or frames, to an event over time" (Meraz & Papacharissi, 2013, p. 144). That said, comparative explorations between the institutional actors and the online public audience have been sparse. In particular, proximity is an important criterion for newsworthiness that affects media institutional framing of global crisis including terrorism (Nossek & Berkowitz, 2006): Understanding how proximity as a news value influences audience's spontaneous frames as well as institutional frames will thus help expand knowledge on networked framing of terrorism.

Terrorism News and Proximity

Proximity is a particularly critical element in deciding to what extent and in which way foreign news is covered (de Vreese, Peter, & Semetko, 2001). For example, Schaefer (2003) examined local, national, and international coverage of two terrorist attacks—the U.S. embassy bombing in Kenya and Tanzania, and the 9/11 attack in the United States—and found prevalence of the "local angle" and "domestication" of distant news by both African and American press (p. 103). More recently, a comparative review of 137 international terrorism news stories covered by media in China and the United States revealed Chinese newspapers' frugal coverage and social value-oriented news framing, contrary to the more prevalent politicized framing in the United States, due to the political distance China maintains with most foreign terrorism events (Zhang, Shoemaker, & Wang, 2013).

In this sense, news proximity—nearness of an individual/entity to a news event itself and/or the subject involved in the event—should affect the ways in which terrorism news is developed. For example, local newspapers of the affected city tend to have not only more coverage but also more action-oriented, concrete depictions of the situation than geographically distant newspapers (Schaefer, 2003); and national newspapers often "domesticate" international attacks by favoring domestic sources and highlighting the impact of the event on their own citizens and government policies (Gerhards & Schafer, 2014; Schaefer, 2003). A comparative study of terrorism news between the United Kingdom and the United States media suggests that different news frames reflect discrepant international relations and foreign policies that each country main-tained (Papacharissi & Oliveira, 2008). Coverage of terrorist attacks among the major TV channels in four countries (CNN, Al-Jazeera, BBC, and ARD) also showed that CNN and Al-Jazeera stories contained geopolitical conflict-oriented frames, whereas BBC and ARD depoliticized the attacks by framing them as individualized criminal acts against humanity (Gerhards & Schafer, 2014); such framing differences have been found not only in news texts but also in photo-journalism (Fahmy, 2010).

That said, the proximity effect on terrorism news has not always been straightforward due to an unclear definition of the concept. Some studies claim that proximity does not have much impact on differentiating international news practices because of global standardization of terrorism news coverage, which leads to the prevalence of episodic frames, moral outrage, and illegitimacy frames (Gerhards & Schafer, 2014; Schaefer, 2003). In the majority of literature, proximity is simply reduced to geographic distance; in other literature, research-ers emphasize "cultural" proximity without offering an operational definition. For example, news coverage of the Afghan war by English and Arabic media (Fahmy, 2005) or the 1996 and 2002 terrorist events in Israel by U.S. and Israeli newspapers (Nossek & Berkowitz, 2006) have been comparatively studied under the premise that one country's media represents a more culturally proximate case than the other. Such context-contingent references to proximity lack conceptual generalizability and may result in inconsistent findings regarding the role of proximity as a news value. Besides, extant studies have paid little attention to the effects of proximity on *audience's minds*. Further conceptual elaboration is needed to engender generalizable knowledge on the impact of proximity as a news value on audience as well as journalistic frames of terrorism.

Construal-Level Theory: Three Axes of Proximity

To conceptualize proximity in a generalizable and operational way, we introduce CLT. CLT explains the effects of proximity on human cognitive processing. Researchers were originally motivated to study it to understand how people thought of the future and made plans for it (Trope & Liberman, 2010). It proposes that individuals use more abstract mental representations—high con-struals—when an object is perceived as distant from self, thus focusing on the *why*. Conversely, individuals use more concrete representations or details—lower construals—when they perceive the object to be closer to them, focusing on the *how* (Trope & Liberman, 2010). For example, when thinking about a terrorist event that occurred far away, a long time ago and to a group that is different from self, one would likely think of social and political reasons that address why it happened, but closer to the place or date of the event, one may likely focus on specific issues such as how to locate missing families, find the perpetrators, steps for reassuring the affected community and other logistical concerns. According to CLT, the perception of distance can dilute individual differences and situa-tional uniqueness, and even intervene in the process of social categorization and stereotyping.

Therefore, proximity is the central concept underlying CLT. Trope and Liberman (2010) pointed out that proximity is a multifaceted concept beyond geographic propinquity, referred to as "psychological distance" (p. 440). Psychological distance is perceived as a combination of social, geographic, and temporal proximities (Stephan, Liberman, & Trope, 2010). First, *geographic proximity* relates to the physical distance of an individual from the place where the event occurs. Experiments showed that respondents used generic language and words when thinking or talking about events that took place in a faraway location. In contrast, when asked to think about the same scenario closer to where they lived, respondents used more specific vocabularies that described tangible and concrete actions (Fujita, Henderson, Eng, Trope, & Liberman, 2006). Henderson (2009) also contended that judging someone's action either as an extension of his or her inherent trait or as circumstantial depends on the geographic and social proximities to the person. For example, individuals tend to evaluate misdeeds more harshly when they occur in a geographically distant place or are conducted by someone outside one's own social boundary compared to proximate misdeeds (Eyal, Liberman, & Trope, 2008). In a similar vein, people tend to perceive distant others as a homogeneous collective unit and their behaviors as representative of the group as opposed to perceiving close others' behaviors as a mark of individuality (Henderson, 2009).

Second, *social proximity* refers to how close one perceives another person as an individual or member of a group (Nan, 2007). An individual feels socially closest to strong ties such as family and friends, followed by weak ties such as colleagues and neighbors and members of an imagined collective community (e.g., compatriots), and socially distant to outsiders who do not belong to his or her community. Thus, when talking about individuals who are socially prox-imate, a person is more likely to use specific, situational factors versus the dispositional qualities they attribute to individuals who belong to "out-groups" (Trope & Liberman, 2010). For example, if you or a friend arrived late to a meeting, you are likely to blame the circumstances at the time, such as traffic (low construal), but if a third person arrived late to a meeting, you are more likely to blame the person's attitude, such as he or she is not punctual and/or does not value your time (high construal).

Last, *temporal proximity* refers to the time when an event occurs—past or future—and its influence on how individuals think about the event (Trope & Liberman, 2010). Distant futures—events that are scheduled to occur in months or even years—are talked about in abstract and general terms, although near future events—scheduled for the same week or the next—are addressed in specific and concrete details. For example, Carter and Sanna (2008) conducted an experiment in which they asked respondents to imagine meeting a prospective employer for a position and list the qualifications that made them suitable for the job. The authors found that respondents were more likely to use indirect self-presentation statements such as highlighting connections to others, talking posi-tively about their group associations ("I graduated from Harvard"), if they imagined the meeting will take place in 3 months—higher construals. On the contrary, those who thought they would meet the employer immediately were more likely to use direct self-presentation statements such as specific qualities and self-attributes ("I have published solo-authored papers in top peer-reviewed journals") to describe their suitability for the job—lower construals.

RESEARCH QUESTION AND HYPOTHESES

The tenet of CLT that three axes of proximity—social, geographic, and temporal —are important in shaping human perceptions and judgments has resonance with framing literature. Framing contends that proximity as a news value influences the ways in which an event is perceived. For example, previous framing studies have shown that proximity is directly related with resource availability and gatekeeping practices and thus influence journalists' choices of frames (Papacharissi & Oliveira, 2008; Weimann & Brosius, 1991). However, this view has not sufficiently addressed what influences the general public's choice of frames.

CLT-based understanding is advantageous in this sense because it focuses on universal human cognitive reactions to psychological distance, an amalgam of perceived geographic, temporal, and social proximities (Trope & Liberman, 2010). Audience framing could reveal stronger proximity effects than media institutional framing because global standardization of news reporting style can result in uniform institutional frames (Gerhards & Schafer, 2014; Schaefer, 2003), whereas audiences may not internalize journalism norms.

In their analysis of how dominant frames emerged in Twitter discourse around an event, Meraz and Papacharissi (2013) found that even though tweets by news organizations and journalists were among the first to be widely shared and retweeted by audiences, other individuals who were providing consistent updates and relevant information were also retweeted by the crowds, thus affording both news media entities and influential citizens an "elite" status. Although these elites played an important role in creating and disseminating narratives that defined the dominant frames around the event on Twitter, further understanding is needed in terms of the differences or similarities between elite and nonelite actors (Meraz & Papacharissi, 2013). Therefore, we ask whether nonelite audiences will be more susceptible to proximity effects on their framing of terrorism than media institutional actors.

RQ1: Do geographic, temporal, and social proximities influence more on audience frames than on media-institutional frames in Twitter?

143

Linking CLT to framing research allows an examination of the effects of psychological distance on the level of "abstractness" in terrorism discourse. Two existing frames are pertinent to address the level of abstractness: *episodic-thematic frames* (Iyengar, 1994) and *space frame* (Chyi & McCombs, 2004).

Episodic-thematic frames distinguish between two ways of storytelling: whether the news narrative provides concrete information on how the event occurred and evolved (*episodic* frame) or more abstract, generic views on the event (*thematic* frame). We posit a series of hypotheses regarding the effect of each dimension of proximity on the use of the *thematic* frame (as opposed to an *episodic* frame). In line with CLT's propositions, we hypothesize that for a user, the more distant the terrorist attack, or an entity related to the attack, the more likely he or she will use a *thematic* frame.

> H1: *Thematic* frames will be deployed more frequently than *episodic* frames if (a) the terrorist attack occurs in a more distant physical location (i.e., geographic proximity), (b) the message posting time is further from the moment of the attack (i.e., temporal proximity by posting time), (c) the referenced temporal perspective is further from the moment of the attack (i.e., referenced temporal proximity in message), and (d) the perceived social actors in the context of the attack are outside their social boundary (i.e., social proximity).

Per CLT, psychological distance influences a person's decision to apply individual-oriented or social identity-based criteria when judging others (Henderson, 2009). This resonates with Chyi and McCombs's (2004) contention that temporal proximity influences the breadth of coverage in the cycle of news life. That is, at the onset of the event (close in time), news coverage tends to emphasize individual-oriented storytelling (low construal). However, as days go by (temporally further from the event), news coverage begins to focus on topics related to social groups, community, region, national, and even geopolitical issues (higher construal). This shift from micro- to macrolevels of framing is indicative of the use of higher construals as the event moves further away in time. Thus, we hypothesize that not only temporal proximity but also other proximity factors influence the spatial orientation of frames, such that

> H2: A higher level of *spatial* frame will be deployed in news messages as (a) the terrorist attack occurs in a more distant location, (b) the message posting time is farther from the moment of the attack, (c) the referenced temporal perspective is farther from the moment of the attack, and (d) perceived social actors in the context of the attack are outside the users' social boundary.

METHODS

Data Collection

Twitter data were collected for about two weeks in the immediate aftermath of BMB13 on April 15, 2013, and BAB16 on March 22, 2016. The BMB13 data collection began a few hours after the incident and continued until April 29, 2013. Twitter StreamingAPI was used to collect tweets with three search keywords: *#BostonMarathon, Boston Marathon*, and *Boston*. Two sessions of data collection were held each day. The total number of tweets collated was 23,300. The BAB16 data were collected using NodeXL-Basic (2014) with the search keyword *Brussels*. NodeXL-Basic (http://www.smrfoun dation.org/nodexl/) uses StreamingAPI and permits the collection of 2,000 tweets every few minutes. The collection began a few hours after the incident on March 22 and continued until April 3, 2016. Two sessions of data collection were held each day. Each collection session lasted for 2 to 3 hours. The data set contained 80,933 tweets.

We organized the data from each data set into a subset by selecting messages that originated in the Western European region or the United States via the tweets' geolocation. As a result, 14,054 tweets from BMB13 and 27,433 from BAB16 were retained in this subset.

Sampling and Geographic Proximity

Geography-based stratified, nonprobabilistic sampling of 4,000 tweets (2,000 for each event) was performed on the chronologically ordered data set for a balanced sample that represents various geographic proximities while also minimizing temporal bias.

Geographic proximity was defined on three levels of distance: For BMB13, we defined the East Coast area in the United States to be the most proximate to the attack (the region in which Boston is located), and the rest of the United States was the second proximate (national); Western European countries were considered the farthest from the place of the incident (international). For BAB16, on the other hand, Brussels/Belgium was defined as the most proximate location to the attack, and other Western European countries were the second proximate (nearby European Union nations); the United States was considered the farthest location (international).

Stratified random sampling of 2,000 from BMB13 resulted in the selection of 867 tweets from the East Coast, 896 from the rest of the United States, and 237 from Europe. When sampling the East Coast tweets, we oversampled the most proximate tweets. All tweets associated with mid-Atlantic areas (New England states) were included for the best representation of local tweets.

Regarding BAB16, only 952 tweets (3.4%) originated from Brussels/ Belgium, suggesting that random sampling would result in very few tweets from Brussels. Accordingly, we separately generated 500 randomized samples from this pool and then performed stratified random sampling of 1,500 tweets from the pool of those messages that originated from the United States and the rest of Western Europe. As a result, 1,026 U.S. tweets, 474 Western European tweets, and 500 Brussels/Belgium tweets were included in the final sample presented here. Although this sampling strategy is not perfectly probabilistic, it corresponds with our goal of exploring the nature of tweets from different geographic locations in a balanced way.

Content Analysis: Variables

Data embedded in the tweets were of three subcategories: user profiles, tweet messages, and the news resources (URL). Four graduate students were trained multiple times to code the relevant variables in this content analysis. Based on the intercoder reliability test of 400 tweets, coders with the highest agreement rate were paired together to complete the coding for the entire data set. Although all reliability scores fell into the acceptable range based on Fleiss's (1971) criteria, Cohen's kappa scores for some variables had somewhat low kappa values largely due to the unbalanced category sample size. The intercoder agreement rates for these variables, however, were high, ranging between 82% and 97%, thus confirming the acceptable coding framework. Table 1 presents more details of the coding framework and reliability test results.

User Profile. A binary category of "personal user" was used to identify audiences. Drawn from Kwon, Oh, Manish and Rao (2012), a profile was *not* considered as a personal user account if it was closely reflective of journalism, media actors, or institutional news sources or specifically meeting any of the five criteria specified in Table 1.

Frames. Frames were analyzed based on tweet messages. We used Iyengar's (1994) original conceptualization of *episodic-thematic* frames in this study. For the *space* frame, we modified Chyi and McCombs's (2004) original five-level variable (individual, community, regional, societal, and international) into a four-level variable due to the rare occurrence of the regional level frame in our sample. The *space* frame was treated as an interval variable (individual = 1, international = 4).

Social Proximity. Snefjella and Kuperman's (2015) conceptualization of social proximity was adopted to identify the different markers of this variable in a tweet message. The authors defined it as an individual's "willingness to establish social contacts with representatives of a racial, ethnic, socioeconomic, occupational, or other social group" (p. 1455). The modified *social proximity* variable include four categories of social references: strong ties (family and

TABLE 1
Coding Framework and Intercoder Reliability

Variable Description (Example)	%	CK
Personal profile: belongs to laymen/audience if *NOT* falling into any of the following:	.87	.72
(a) politicians or government personnel; (b) clear indication of the primary career in journalism, such as journalist, anchor, reporter, news editor, radio host, etc., or affiliation with a specific media organization or media-centric advocacy institution; (c) strict, career-only description with clear identification with certain organization without providing any personal information; (d) having only a series of URLs or hashtags; (e) group/organizational account ("*News talk 980 CKNW. Vancouver's news. Vancouver's talk.*")		
Iyengar's frame: The message frames the event . . .	.91	.63
(1) Episodic = in specific terms describing the place, event or person involved ("*A top-ranking rep. says more arrests possible in #Bostonbombing.*") (2) Thematic = in political/social/cultural context, or in an abstract manner ("*We live in a sad, mad, tragic world. #prayersforBoston*")		
Space frame: The message is framed by highlighting. . .	.84	.61
(1) Individuals directly involved in the event ("*FBI interviewed Boston bombing suspect in 2011.*") (2) Community, city, region ("*Cell phones in Boston are out of service.*") (3) The whole society or nation ("*After the arrest of Boston Bomber I think American security services are good and swift.*") (4) International repercussions (e.g., "*Christians are in Jail in #Iran for being Christian. Why aren't American Muslims speaking out?*")		
Social distance: The message includes any reference to. . .		
(1) Strong ties: family and friends ("*617–635-4500 is the number concerned families can call if you can't find family/runners*")	.97	.91
(2) Compatriot: fellow citizens, or public servants ("*Come on. Our country is better than this MT @PatchTweet Muslim woman assaulted, blames #BostonMarathon bombing*")	.83	.70
(3) Empathic strangers: victims or affected who are sympathized ("*An innocent man running for charity and this happens. It's sickening.*")	.89	.84
(4) Outsiders: otherness such as immigrant, foreigner, alien ("*To immigrants. If you don't want to change your life style to that of the west, don't come.*")	.93	.77
News resource: Whether the hyperlink is broadcast websites (TV/radio); newspaper website; digital news or information sites; social media	.82	.75
Referenced temporal proximity: The message talks about . . .	.84	.5
(1) Present issues/happening ("*Sending prayers to those families hurt.*") (2) Past ("*Most American Muslims supported the insulting mosque at ground zero, a monument the ideology inspired the 911.*") (3) Future ("*#Bostonmarathon probe and #bigdata use hints at the future.*")		

Note. CK = Cohen's kappa scores.

friends), compatriots, empathic strangers (victims of the terror attack toward whom people express sympathy on social media), and outsiders (visitors and foreigners).[1] We treated each one as a categorical variable because a single message could refer to multiple categories. Among them, "outsiders" was the exclusionary reference, whereas other categories commonly pertained to in-group members despite varied relational strengths.

Temporal Proximity. Two kinds of temporal proximity were examined. First, the posting time of tweet messages was used. Studies suggest that event-oriented news coverage tends to occur the most during the first couple of days immediately after the incident, then drastically decrease about a week later (Chyi & McCombs, 2004). The time span of Twitter activity is even shorter as the vast majority of tweeting/retweeting occurs within a few hours of the event, and very little occurring beyond 2 days (Kwak, Lee, Park, & Moon, 2010). Thus, we split the chronologically ordered data into three time windows: 1 = Day 1–2, 2 = Day 3–7, 3 = Day 8 and the rest. Second, the temporal reference in tweet messages was used as another dimension of temporal proximity. For this, we adopted Chyi and McCombs's (2004) time frame composed of present, past, and future references.

Control Variables. First, referencing a news resource could affect audiences' frame construction, as sharing reported stories could perpetuate media frames. Accordingly, we controlled for types of *news resource reference* (i.e., URLs embedded in tweets) conceptualized as per Kwon et al. (2012): TV/radio originated, press originated, social media, and other digital resources (such as online journalism, database, academic resource, etc.). Second, we considered the use of *hashtags*, as these are shared markers reveal the ways in which the event is portrayed.

RESULTS

Descriptive Analyses

Descriptive statistics suggest that BMB13 and BAB16 are similar in some ways. The majority of tweets in both data sets were sent from nonmedia personnel or audience (61.45% for BMB13, 57.85% for BAB16), were heavily episodic (89.46% for BMB13, 69.06% for BAB16), and mentioned present issues as opposed to conveying future or past points of view (79.59% for BMB13, 94.72% for BAB16). Among social proximity categories, both data sets showed relatively frequent reference to "compatriots" (17.5% for BMB13, 23.24% for BAB16) and "empathic strangers" (16.20% for BMB13, 13.22% for BAB16).

---

[1] The original categories included "weak ties" composed of neighbors and coworkers. This category was excluded due to rare occurrences.

148

Meanwhile, differences emerged in terms of the *space* frame and reference to the "outsiders" category. In the BMB13 tweets, individual story-oriented framing was predominant (77.61%). Community/regional level framing (42.98%), however, was as frequent as individual level (37.79%) in the BAB16 set. Moreover, BAB16 contained more references to outsiders (11.85%) compared to BMB13 (6.65%) (Figure 1).

Before investigating the relative effects of proximity variables, we examined mean or frequency difference of frames, using chi-square and Kruskal-Wallis rank analysis of variance tests. The frame differences were found consistently in terms of geographic proximity and posting time. The referenced temporal proximity also resulted different frames except the thematic/episodic frames in BAB16. Audiences and media actors' framing were different in BMB13 but not in BAB16. The test results are summarized in Table 2.

Proximity Effects on Thematic/Episodic Frames

Logistic regression modeling was performed to examine proximity effects on the use of Iyengar's frames (thematic = 1). As shown in Table 3, geographic proximity was significant only for BAB16; the farther the event, the more thematic the frame. The U. S. media ($b = .51$, odds ratio $[OR] = 1.67$, $p < .05$) and audiences ($b = .69$, $OR = 1.99$,

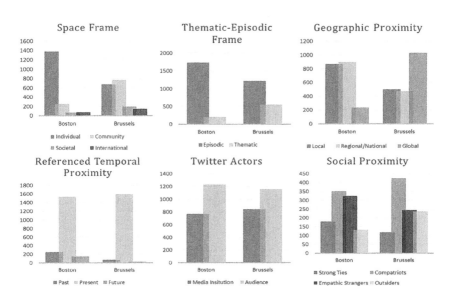

FIGURE 1 Frequency of +.

TABLE 2
Frame Differences by Proximity Variables

	Boston Marathon Bombing 2013		Brussels Airport Bombing 2016	
	Space[a]	Iyengar's[b]	Space[a]	Iyengar's[b]
Geographic	$\chi^2(2) = 14.201$***	$\chi^2(2) = 7.276$	$\chi^2(2) = 23.761$***	$\chi^2(2) = 15.154$**
Temporal: Referenced	$\chi^2(2) = 13.578$**	$\chi^2(2) = 16.885$***	$\chi^2(2) = 13.712$**	$\chi^2(2) = 5.120$
Temporal: Posting time	$\chi^2(2) = 3.775$	$\chi^2(2) = 23.317$***	$\chi^2(2) = 7.769$*	$\chi^2(2) = 8.711$*
Actors (Audiences vs. Media/Organizations)	$\chi^2(1) = 5.743$*	$\chi^2(1) = 18.139$***	$\chi^2(1) = 3.539$	$\chi^2(1) = 3.042$
Social distance				
Strong ties	$\chi^2(2) = 28.001$***	$\chi^2(2) = .974$	$\chi^2(2) = 47.962$***	$\chi^2(2) = 33.585$***
Compatriots	$\chi^2(2) = 0.005$	$\chi^2(2) = 1.101$	$\chi^2(2) = 34.689$***	$\chi^2(2) = 0.236$
Empathic strangers	$\chi^2(2) = 13.774$***	$\chi^2(2) = 11.439$**	$\chi^2(2) = 85.677$***	$\chi^2(2) = 38.878$***
Outsiders	$\chi^2(2) = 207.267$***	$\chi^2(2) = 287.959$***	$\chi^2(2) = 14.034$***	$\chi^2(2) = 64.249$***

[a]Kruskal-Wallis test (nonparametric one-way analysis of variance) was used due to the unequal population variances. [b]Chi-square test.
*$p < .05$. **$p < .01$. ***$p < .001$.

$p < .01$) were 1.67 times and 1.99 times more likely, respectively, to use a thematic frame to talk about BAB16 than domestic media and audiences in Belgium.

Social proximity effect was consistent with the hypotheses. The outsiders category increased the likelihood of thematic framing more than 13 times in BMB13 ($b = 2.58$, $OR = 13.19$, $p < .001$, for media; $b = 2.61$, $OR = 13.65$, $p < .001$, for audience) and almost three times in BAB16 ($b = 1.10$, $OR = 2.99$, $p < .001$, for media; $b = 1.07$, $OR = 2.91$, $p < .001$, for audience). Also, empathic strangers and strong ties in some models decreased the likelihood of a thematic frame—and thus increased the use of episodic frame. For example, references to empathic strangers decreased the chance of the audience using the thematic frame in BMB13 by 64% ($b = -1.02$, $OR = .36$, $p < .01$) and by 51% for BAB16 ($b = -.71$, $OR = .49$, $p < .01$).

In regards to temporal proximity effects, only audience-related tweets showed thematic framing ($b = .77$, $OR = 2.15$, $p < .05$ for BMB13; $b = .47$, $OR = 1.60$, $p < .05$, for BMB16) with decreasing temporal proximity between the event and tweet posting time (i.e., 1 week or later). Another significant finding was that the referenced temporal proximity showed the *opposite* pattern than hypothesized. Specifically, the audience frame in BMB13 showed a negative effect of past and future references, suggesting that decreasing temporal proximity (moving further away from the time of the event) was associated with the episodic frame ($b = -1.06$, $OR = .35$, $p < .01$, for past reference; $b = -1.54$, $OR = .21$, $p < .05$, for future reference).

Proximity Effects on Space Frame

Ordinary least squares regression modeling was performed to examine the *space* frame. Significant geographic proximity effect was found only in the audience frames of BMB13, suggesting that European audiences were more likely to use a higher level of space frame (Table 4) to talk about the Boston Bombing, compared to local audiences on the East Coast of the United States ($\beta = .09$, $t = 2.99$, $p < .01$). In all other models, geographic proximity effect was not significant.

Among social proximity categories, strong ties were consistently associated with the lower level (more individual oriented) space frame for both, audiences ($\beta = -.09$, $t = 2.22$, $p < .05$, in BMB13; $\beta = -.12$, $t = 3.52$, $p < .001$, in BAB16), and news media ($\beta = -.09$, $t = 2.89$, $p < .01$, in BMB13; $\beta = -.14$, $t = 3.64$, $p < .001$, in BAB16). Reference to empathic strangers in BAB16 was also associated with the lower space frame ($\beta = -.18$, $t = 4.69$, $p < .001$, for media; $\beta = -.17$, $t = 4.92$, $p < .001$, for audience). In contrast, reference to outsiders in BMB13 was associated with the higher level of space frame, and the explained variance was large ($\beta = .53$, $t = 16.22$, $p < .001$, for media; $\beta = .50$, $t = 18.84$, $p < .001$, for audience).

The posting time effect was not significant. Reference to future was significant for news media's space frame in BMB13; the effect was opposite to our

TABLE 3
Proximity Effects on Terrorism Framing in Social Media: Iyengar's Frame (Thematic = 1)

| | Boston Marathon Bombing 2013 | | | | | | Brussels Airport Bombing 2016 | | | | | |
| | Media[a] | | | Audience[b] | | | Media[c] | | | Audience[d] | | |
	b	SE	OR	b	SE	OR	b	SE	OR	b	SE	OR
Geographic 1	.42	.34	1.52	.09	.23	1.09	.33	.28	1.40	.86	.26	2.37**
2	.70	.62	2.00	.38	.30	1.47	.51	.24	1.67**	.69	.23	1.99**
Temporal: Posting time 3–7	-.41	.60	.66	-.21	.33	.81	.25	.23	1.29	.12	.18	1.13
8+	.39	.61	1.48	.77	.34	2.15*	.02	.25	1.02	.47	.22	1.60*
Temporal: Past	-.80	.56	.45	-1.06	.39	.35*	-.23	.61	.80	-.41	.51	.66
Referenced: Future	-1.13	.67	.32	-1.54	.78	.21*	-.50	1.13	.61	—	—	—
Social distance ST	.36	.50	1.43	-.17	.40	.84	-2.69	1.03	.07**	-1.52	.45	.22**
CMP	-.05	.42	.95	.49	.25	1.64	.17	.20	1.18	-.13	.18	.88
ES	-.81	.53	.45	-1.02	.37	.36**	-1.15	.38	.32**	-.71	.27	.49**
OS	2.58	.42	13.19***	2.61	.26	13.65***	1.10	.27	2.99***	1.07	.22	2.91***
Hashtags	.07	.11	1.08	.16	.06	1.17	.20	.09	1.22*	.18	.06	1.20**
News resources Broad	-1.12	.65	.33	-.83	.52	.44	-.64	.31	.53*	-1.04	.32	.35**
Press	.08	.51	1.08	-.98	.55	.38	.28	.27	1.32	.45	.24	1.58
Digital	.10	.39	1.10	-.29	.27	.75	-.07	.24	.94	-.52	.22	.59*
SM	-1.78	1.09	.17	-1.46	.39	.23***	-.35	.42	.70	-.67	.32	.51*
Model	$\chi^2(15) = 68.63^{***}$			$\chi^2(15) = 191.34^{***}$			$\chi^2(15) = 82.80^{***}$			$\chi^2(14) = 114.34^{***}$		
	Pseudo-R^2 = .184			Pseudo-R^2 = .214			Pseudo-R^2 = .107			Pseudo-R^2 = .105		

Note. OR = odds ratio; Geographic 1 = the United States and 2 = Europe in Boston Marathon bombing 2013, 1 = Europe and 2 = the United States in Brussels Airport bombing 2016 (Reference = the East Coast in Boston Marathon bombing 2013 and Belgium in Brussels Airport bombing 2016); ST = strong ties; CMP = compatriots; ES = empathic strangers; OS = outsiders; Broad = broadcast (TV/radio); SM = social media.
[a]N = 748. [b]N = 1,175. [c]N = 643. [d]N = 846.
*p < .05. **p < .01. ***p < .001.

TABLE 4
Proximity Effects on Terrorism Framing in Social Media: Space Frame

| | Boston Marathon Bombing 2013 | | | | | | Brussels Airport Bombing 2016 | | | | | |
| | Media[a] | | | Audience[b] | | | Media[c] | | | Audience[d] | | |
	β	SE	t	β	SE	t	β	SE	t	β	SE	t
Geographic												
1	-.02	.05	-.57	.02	.04	.59	-.08	.10	-1.63	.03	.10	.59
2	.00	.10	.12	.09	.06	2.99**	-.03	.08	-.61	-.08	.08	-1.77
Temporal: Posting time												
3–7	.01	.09	.20	.02	.06	.52	-.02	.08	-.43	-.01	.07	-.28
8+	.00	.09	-.06	.00	.07	-.04	.05	.08	1.01	.00	.09	.08
Temporal: Referenced												
Past	-.05	.06	-1.43	-.02	.06	-.77	-.04	.20	-1.07	-.02	.18	-.56
Future	-.11	.07	-2.95**	.00	.11	.08	-.01	.34	-.16	-.03	.51	-.78
Social distance												
ST	-.09	.08	-2.89**	-.06	.07	-2.22*	-.14	.13	-3.64***	-.12	.12	-3.52***
CMP	-.02	.06	-.48	.01	.05	.34	-.07	.07	-1.79	-.13	.07	-3.97***
ES	-.09	.06	-2.55*	-.02	.05	-.90	-.18	.10	-4.69***	-.17	.09	-4.92***
OS	.53	.10	16.22***	.50	.07	18.84***	.06	.10	1.45	.05	.09	1.41
Hashtags	-.01	.02	-.27	.04	.01	1.64	.10	.03	2.64**	.06	.02	1.77
News resources												
Broad	-.07	.07	-1.93	-.04	.09	-1.39	-.10	.10	-2.35*	-.11	.11	-3.13**
Press	-.01	.08	-.25	-.01	.09	-.53	-.04	.10	-.86	-.09	.10	-2.42*
Digital	-.06	.06	-1.68	-.03	.06	-1.06	-.07	.09	-1.46	-.09	.08	-2.44*
SM	.00	.09	-.02	.09	.06	3.29**	-.02	.15	-.51	.00	.12	-.02
Model	F(15, 669) = 20.92***			F(15, 1064) = 28.85***			F(15, 625) = 5.16***			F(15, 825) = 6.15***		
	Adj.R^2 = .304			Adj.R^2 = .279			Adj.R^2 = .089			Adj.R^2 = .084		

Note. Geographic 1 = the United States and 2 = Europe in Boston Marathon bombing 2013, 1 = Europe and 2 = the United States in Brussels Airport bombing 2016 (Reference = the East Coast in Boston Marathon bombing 2013 and Belgium in Brussels Airport bombing 2016); ST = strong ties; CMP = compatriots; ES = empathic strangers; OS = outsiders; Broad = broadcast (TV/radio); SM = social media.
[a]N = 685. [b]N = 1,080. [c]N = 641. [d]N = 841.
*p < .05. **p < .01. ***p < .001.

hypothesis, suggesting that future reference was associated with lower level of space frame ($\beta = -.11$, $t = 2.95$, $p < .01$).

News Resource Effects

The effects of hyperlinking to news sources are noteworthy, although they were not hypothesized. In some models, linking to news sources was significant, most of which was negatively associated. That is, compared to those not including any external news links, tweets that included a news media url were associated with the episodic frame and lower level space frame. The exception, however, was the effect of social media reference on audience's space frame in BMB13, which showed an increase in the thematic frame ($\beta = .09$, $t = 3.29$, $p < .01$). Also, the hyperlinking effect appeared more frequently in audience framing than in media institutional framing.

DISCUSSIONS AND CONCLUSION

By comparatively exploring media/institutional frames and audience frames on Twitter, this study builds on discussions around social media–afforded networked framing in the context of terrorism (Meraz & Papacharissi, 2013). This study also delved into the role of news "proximity" in both audience and media frames of terrorism. A unique contribution of this study was to elaborate three generalizable and operational dimensions of proximities drawn from CLT—spatial, social, and temporal proximity.

By applying CLT, we clarified the notion of proximity and examined its impact on terrorism framing as a universal human cognitive mechanism. Framing research focuses on how news media portrays an event. By including audience framing in the picture, analyzed through CLT, this article adds to communication scholars' understanding of audience discourse and framing of a dynamic news event. Audiences' comments and tweets on social media influenced news coverage, especially in the aftermath of the Boston Bombing (Ziv, 2015), and therefore it becomes imperative to analyze people's cognitive response and subsequent information sharing on Twitter during a crisis such as a terrorist attack.

In addition, this study shows interdisciplinary value by adding to audience-centered crisis management scholarship that focuses on how publics respond to crisis information received through various media channels, including social. This information is often a cocreation between people and official sources and an examination of how people may influence one another versus how institutions may influence them becomes important (Liu, Fraustino, & Jin, 2016). Interestingly, a study of Arab audience tweets by Alkazemi, Fahmy, and Wanta

(2017) also suggests that audiences on social media may attempt to influence media coverage by posting nonobjective messages of an issue in which they are personally vested. This line of research becomes even more important as journalists themselves look to social media for information as they wrestle with cuts in manpower and financial resources (Alkazemi et al., 2017).

Proximity Effects on Frames

Results of the entire sample revealed that the tweets were predominantly based on the "present" frame, rather than the past or future, and included concrete, detailed accounts of the event (i.e., episodic) rather than abstract ideas. These findings are unsurprising considering that instantaneousness and personalized information sharing are the defining characteristics of Twitter (Oh, Agrawal, & Rao, 2013). These findings also are in line with previous terrorism framing research.

The two sets of hypotheses were posited such that temporal, geographical, and social distances will produce more high construal frames. Although not all hypothesized relationships were significant, overall trends—where significant —were consistent with CLT's proposition that greater distance engenders more abstract storytelling in the form of a *thematic* frame or a higher order *space* frame. Among proximity effects, social proximity was particularly noteworthy: The significant association between the reference to outsiders (i.e., socially distant others) and *thematic* frames was observed across the different news events (BMB13 and BAB16) and user types (i.e., media institutions or audiences). Such findings fall in line with the premise of CLT and imply the potential for social categorization and stereotyping of outsiders constructed by high construal discourse. In our sample, the nontrivial portion of tweets that referred to outsiders were about immigrants, refugees, or religious groups, and users portrayed these groups as "them" versus "we." An audience tweet exemplifies this observation: "To immigrants. If you don't want to change your lifestyle to that of the west, don't come. If you don't like it here, go back." One possibility is the media's *thematic* account could inadvertently spill over to audience frames in a manner that reinforces their social-categorical view of these populations. We did not test the interaction effect between media's thematic frame and references to outsiders as framed in the audience tweets and recommend further research in this direction.

Distant temporal reference was associated with more episodic and lower level space frames, particularly in BMB13. This finding seems contradictory to previous framing research (e.g., Kwon & Moon, 2009; Chyi & McCombs, 2004) and the posited hypothesis. However, a closer look at the data suggests that a majority of tweets with a past or future reference in our sample were actually about the *short-term* past and future, extending only a few days. Therefore, those tweets continued to talk about "what has been done," "what is going to be done," and "anomalies of terrorists' personalities," as opposed to "why it happens." Audiences continue to focus on specific details when they discuss the immediate past and future, such as information about the individual attackers or forthcoming official statements on terrorists. Such narratives align with previous research that the American media tend to respond to a domestic terror attack by dedicating much coverage to the perpetrators (Kwon & Moon, 2009; Morin, 2016).

Comparison Between Media Institutional and Audience Frames

In response to RQ1, results suggest that overall, proximity effects were more prominent on audience frames than media institutional frames. Most proximity variables were significantly associated with the audience's use of the *thematic* frame in both events, whereas the number of significant variables reduced when media institutional samples were modeled.

We conclude that media and audience frames show a great deal of similarities but do not always converge on Twitter. The episodic frame especially was more frequent in media institutional tweets than audiences' tweets. This is possibly because media and institutional personnel get direct access to attack-related news sources and thus perceive closer psychological distance to the terror event than the general publics. Also, journalistic norms such as objectivity and accuracy could lead to a more concrete, episodic description of events.

Another related observation is that audiences were more prone to hyperlinking information received on social media than media actors. This tendency affects audiences' adoption of the broader spatial frame in terrorism discourse during BMB13. Audiences may be more likely to use alternative resources than journalists, who usually cite institutional sources and official spokespeople. Also, social media is a global and transborder platform. Majority of audiences during BMB13 were the American public, and it is unsurprising that they would choose social media resources that primarily use the English language to look for global views on the terrorism event.

Differences Between the Two Events

Despite some consistent patterns with regard to proximity effects and audience versus media institutional frames in the two samples, some differences emerged. Tweets during BMB13 focused on individual-oriented storytelling and episodic frames, whereas tweets during BAB16 included more of the community/regional frame and the *thematic* frame. Although this difference between the events was not hypothesized, CLT offers a convincing rationale. As a majority of our sample comprised U.S.-originated tweets, it is possible that tweets related to BAB16 (a geographically distant event from the United States) included higher order frames more frequently than BMB13 (a closer event within their own country).

More importantly, geopolitical antecedents surrounding each event may explain this difference even more convincingly. BAB16 occurred 4 months after the Paris terrorist attacks, and police found strong connections between the two events. Concurrently, Western Europe has been struggling with a humanitarian crisis with hundreds of refugees from war-torn Iraq and Syria seeking asylum in various European Union nations. Meanwhile, BMB13 was the first major terrorist attack on American soil in more than a decade since 9/11. Also, it was perpetrated by an American citizen and permanent resident (the Tsarnaev brothers). It is likely, then, the narrative around BAB16 took place within a *thematic* frame and at a higher construal level by incorporating geopolitical perspectives into the terrorism discourse.

That said, inconsistent results between the two events could limit the generalizability of our findings unless geopolitical factors were put into consideration. We recommend future research attend to contextual variables to minimize unknown variances. Another limitation of this study is the unequal data collection and sampling methods between the two events. We are aware of the possibility of sampling bias on the results. Despite the limitation, this research builds on terrorism scholarship and social media research by showing how temporal, social, and physical proximities influence audience and media institutional framing of terrorism. Terrorism news in social media spreads rapidly and efficiently, well beyond domestic audiences. With many people relying on social media for their news, social media framing research may help advance our knowledge about the ways in which terrorism is constructed by publics and institutions in a fast-paced, contemporary information environment.

ACKNOWLEDGMENTS

We are thankful to the reviewers and editors for their constructive comments.

FUNDING

This project was supported by AEJMC Emerging Scholarship 2017.

ORCID

K. Hazel Kwon ⓘ http://orcid.org/0000-0001-7414-6959

REFERENCES

Alkazemi, M., Fahmy, S., & Wanta, W. (2017). The promise to the Arab world: Attributes of U.S. President Obama in Arabic-language tweets. *International Communication Gazette*. Advance online publication. doi:10.1177/1748048517727207.

Carter, S. E., & Sanna, L. J. (2008). It's not just what you say but when you say it: Self presentation and temporal construal. *Journal of Experimental Social Psychology*, *44*(5), 1339–1345. doi:10.1016/j.jesp.2008.03.017

Chyi, H. I., & McCombs, M. (2004). Media salience and the process of framing: Coverage of the Columbine school shootings. *Journalism & Mass Communication Quarterly*, *81*(1), 22–35.

Dalal, N. (2017, January 12). How media fuels our fear of terrorism. *Priceonomics*. Retrieved from https://priceonomics.com/our-fixation-on-terrorism/

Das, E., Bushman, B. J., Bezemer, M. D., Kerkhof, P., & Vermeulen, I. E. (2009). How terrorism news reports increase prejudice against outgroups: A terror management account. *Journal of Experimental Social Psychology*, *45*(3), 453–459.

Davis, D. W., & Silver, B. D. (2004). Civil liberties vs. security: Public opinion in the context of the terrorist attacks on America. *American Journal of Political Science*, *48*(1), 28–46.

de Vreese, C. H., Peter, J., & Semetko, H. A. (2001). Framing politics at the launch of the Euro: A cross-national comparative study of frames in the news. *Political Communication*, *18*(2), 107–122.

Eyal, T., Liberman, N., & Trope, Y. (2008). Judging near and distant virtue and vice. *Journal of Experimental Social Psychology*, *44*(4), 1204–1209.

Fahmy, S. (2005). Emerging alternatives or traditional news gates: Which news sources were used to picture the 9/11 attack and the Afghan War? *Gazette (Leiden, Netherlands)*, *67*(5), 381–398.

Fahmy, S. (2010). Contrasting visual frames of our times: A framing analysis of English-and Arabic-language press coverage of war and terrorism. *International Communication Gazette*, *72*(8), 695–717.

Finseraas, H., & Listhaug, O. (2013). It can happen here: The impact of the Mumbai terror attacks on public opinion in Western Europe. *Public Choice*, *156*(1–2), 213–228.

Fleiss, J. L. (1971). Measuring nominal scale agreement among many raters. *Psychological Bulletin*, *76*(5), 378.

Fujita, K., Henderson, M. D., Eng, J., Trope, Y., & Liberman, N. (2006). Spatial distance and mental construal of social events. *Psychological Science*, *17*(4), 278–282.

Gadarian, S. K. (2010). The politics of threat: How terrorism news shapes foreign policy attitudes. *The Journal of Politics*, *72*(2), 469–483.

Gerhards, J., & Schäfer, M. S. (2014). International terrorism, domestic coverage? How terrorist attacks are presented in the news of CNN, Al Jazeera, the BBC, and ARD. *International Communication Gazette*, *76*(1), 3–26. doi:10.1177/1748048513504158

González-Ibáñez, R., Muresan, S., & Wacholder, N. (2011). Identifying sarcasm in Twitter: A closer look. In *Proceedings of the 49th Annual Meeting of the Association for Computational Linguistics: Human Language Technologies*: *Short Papers - Volume 2* (pp. 581–586). Portland, OR: Association for Computational Linguistics.

Henderson, M. D. (2009). Psychological distance and group judgments: The effect of physical distance on beliefs about common goals. *Personality and Social Psychology Bulletin*, *35*(10), 1330–1341.

Huddy, L., Khatib, N., & Capelos, T. (2002). Trends: Reactions to the terrorist attacks of September 11, 2001. *The Public Opinion Quarterly, 66*(3), 418–450.

Iyengar, S. (1994). *Is anyone responsible? How television frames political issues.* Chicago, IL: University of Chicago Press.

Kwak, H., Lee, C., Park, H., & Moon, S. (2010). What is Twitter, a social network or a news media? In *Proceedings of the 19th International Conference on World Wide Web* (pp. 591–600). New York, NY: ACM.

Kwon, K. H., & Moon, S. I. (2009). The bad guy is one of us: Framing comparison between the US and Korean newspapers and blogs about the Virginia Tech shooting. *Asian Journal of Communication, 19*(3), 270–288.

Kwon, K. H., Oh, O., Manish, A., & Rao, H. R. (2012). Audience gatekeeping in the Twitter service: An investigation of tweets about the 2009 Gaza Conflict. *AIS Transaction on Human-Computer Interaction, 4*(4), 212–229.

Kwon, K. H., Xu, W. W., Wang, H., & Chon, J. (2016). Spatiotemporal diffusion modeling of global mobilization in social media: The case of 2011 Egyptian revolution. *International Journal of Communication, 10*, 73–97.

Landau, S. (2011). *Surveillance or security? The risks posed by new wiretapping technologies.* Cambridge, MA: MIT Press.

Lewis, S. C., & Reese, S. D. (2009). What is the war on terror? Framing through the eyes of journalists. *Journalism & Mass Communication Quarterly, 86*(1), 85–102.

Liu, B. F., Fraustino, J. D., & Jin, Y. (2016). Social media use during disasters: How information form and source influence intended behavioral responses. *Communication Research, 43*(5), 626–646.

McLeod, D. M., & Shah, D. V. (2015). *News frames and national security: Covering big brother.* New York, NY: Cambridge University Press.

Meraz, S., & Papacharissi, Z. (2013). Networked gatekeeping and networked framing on # Egypt. *The International Journal of Press/Politics, 18*(2), 138–166.

Morin, A. (2016). Framing terror: The strategies newspapers use to frame an act as terror or crime. *Journalism & Mass Communication Quarterly.* Advance online publication. doi:10.1177/1077699016660720

Nacos, B. L., & Torres-Reyna, O. (2003). Framing Muslim-Americans before and after 9/11. In P. Norris, M. Kern, & M. Just (Eds.), *Framing terrorism: The news media, the government, and the public* (pp. 133–158). New York, NY: Routledge.

Nan, X. (2007). Social distance, framing and judgment: A construal level perspective. *Human Communication Research, 33*, 489–514.

National Consortium for the Study of Terrorism and Responses to Terrorism. (2015). *American death in terrorist attacks* [Fact sheet]. Retrieved from https://www.start.umd.edu/pubs/START_AmericanTerrorismDeaths_FactSheet_Oct2015.pdf

Nip, J. Y., & Fu, K. W. (2016). Networked framing between source posts and their reposts: An analysis of public opinion on China's microblogs. *Information, Communication & Society, 19*(8), 1127–1149.

Norris, P., Kern, M., & Just, M. (2003). *Framing terrorism: The news media, the government, and the public.* New York, NY: Routledge.

Nossek, H., & Berkowitz, D. (2006). Telling "our" story through news of terrorism: Mythical newswork as journalistic practice in crisis. *Journalism Studies, 7*(5), 691–707.

Oh, O., Agrawal, M., & Rao, H. R. (2013). Community intelligence and social media services: A rumor theoretic analysis of tweets during social crises. *MIS Quarterly, 37*(2), 407–426.

Papacharissi, Z., & Oliveira, M. (2008). News frames terrorism: A comparative analysis of frames employed in terrorism coverage in US and UK newspapers. *The International Journal of Press/Politics, 13*(1), 52–74.

Reese, S. D. (2001). Prologue–framing public life: A bridging model for media research. In S. D. Reese, O.H. Gandy Jr., & A. E. Grants (Eds.), *Framing public life* (pp. 7–31). Mahwah, NJ: Lawrence Erlbaum Associates.

Schaefer, T. M. (2003). Framing the US embassy bombings and September 11 attacks in African and US newspapers. In P. Norris, M. Kern, & M. Just (Eds.), *Framing terrorism: The news media, the government, and the public* (pp. 93–112). New York, NY: Routledge.

Snefjella, B., & Kuperman, V. (2015). Concreteness and psychological distance in natural language use. *Psychological Science, 26*(9), 1449–1460.

Stephan, E., Liberman, N., & Trope, Y. (2010). Politeness and psychological distance: A construal level perspective. *Journal of Personality and Social Psychology, 98*(2), 268–280.

Traugott, M. W., & Brader, T. (2003). Explaining 9/11. In P. Norris, M. Kern, & M. Just (Eds.), *Framing terrorism: The news media, the government, and the public* (pp. 159–179). New York, NY: Routledge.

Trope, Y., & Liberman, N. (2010). Construal-level theory of psychological distance. *Psychological Review, 117*(2), 440–463.

Weimann, G., & Brosius, H. B. (1991). The newsworthiness of international terrorism. *Communication Research, 18*(3), 333–354.

World Values Survey (2015). *1981-2014 longitudinal aggregate version 20150418*. Retrieved from http://www.worldvaluessurvey.org/ WVSDocumentationWVL.jsp

Zhang, D., Shoemaker, P. J., & Wang, X. (2013). Reality and newsworthiness: Press coverage of international terrorism by China and the United States. *Asian Journal of Communication, 23*(5), 449–471.

Ziv, S. (2015, April 15). How social media changed news coverage after the Boston Marathon attack. *Newsweek*. Retrieved from http://www.newsweek.com/how-social-media-changed-reporting-wake-boston-marathon-attack-322416

U.S. News Coverage of Global Terrorist Incidents

Mingxiao Sui

Johanna Dunaway

David Sobek

Mingxiao Sui (Ph.D., Louisiana State University, 2017) is an assistant professor in the Department of Media and Communication at Ferrum College. Her research interests include political communication, race and ethnicity in politics, newsroom diversity, social media and mobile communication, and international communication.

Johanna Dunaway (Ph.D., Rice University, 2006) is an associate professor in the Department of Communication at Texas A&M University. Her research interests include political communication, media coverage of groups, issues, and events, public opinion, political behavior and the changing media landscape.

David Sobek (Ph.D., Penn State University, 2003) is an associate professor in the Department of Political Science at Louisiana State University. His research interests include international conflict.

Andrew Abad (M.M.C., Louisiana State University, 2016) is a graduate of the Manship School of Mass Communication at Lousiana State University. His research interests include political comedy, agenda setting, and political learning through mass media.

Lauren Goodman (M.A., Louisiana State University, 2016) is a graduate of the Manship School of Mass Communication at Lousiana State University. Her research interests include crisis communication, media framing, and priming.

Paromita Saha (M.A., Louisiana State University, 2014) is a graduate of the Manship School of Mass Communication at Lousiana State University. His research interests include journalism norms and routines, media ethics, media diversity, race representation in the media, and political economy of the media.

Andrew Abad, Lauren Goodman, and Paromita Saha

A slew of gruesome executions by terrorist groups in 2014–2015 renewed interest in the public relations strategies of terrorists. As a case in point, the Islamic State group's escalating brutality reflects their efforts as a relatively nascent extremist group to ensure a high and sustained volume of media coverage, especially among Western outlets. But what characteristics of events actually prompt coverage from major U.S. news media? Using a rich data set of terrorist incidents and coverage from six major broadcast and cable U.S. networks, we model coverage of terrorist incidents as a function of event proximity from U.S. soil, target country affinity with the United States, number of total and U.S. casualties, and the characteristics of the terrorist group. Our findings largely corroborate expectations set forth by the literature on norms and routines of journalism and economics of news. When it comes to terrorism, coverage by U.S. major media outlets is largely dependent on proximity to and affinity with the United States, weapons of mass destruction, and the number of global and U.S. casualties.

On February 6, 2017, President Donald J. Trump used his first speech to men and women of the military to accuse the U.S. press of deliberately underreporting terrorism (Gajanan, 2017). This accusation was leveled at news media against a backdrop of recent years intensely populated by global terrorist events; historic levels of elite polarization; increasing hostility between partisans; and a press decimated by intense economic competition, shrinking numbers of international bureaus, eroding ranks of journalists, and rapidly declining media trust. Conditions were ripe for using terrorism as a wedge issue and indicting the press for subverting U.S. interests. Veracity of the specific accusations aside, these conditions sustain a broad trend whereby politicians strategically campaign against the press, not so subtly questioning their credibility and objectivity. These circumstances raise important questions about which terrorist events attract coverage. Although evidence generally concludes that newsmaking is a collective process affected by the norms and routines of journalism (Cook, 2005), it is important to take stock of U.S. news media coverage of terrorism in contemporary media and political contexts.

The proliferation of round-the-clock news channels has expanded the spectacle of global terrorism (Hoskins & O'Loughlin, 2007). Even in the digital era, television news remains the most widely used news platform in America; cable news audiences are still growing (Mitchell, Gottfried, Barthel, & Shearer, 2016).

Terrorist groups operate on the premise that more blood and violence means more Western mainstream media coverage (Jetter, 2015). News media use these gratuitous and graphic videos to highlight the impending threat of these groups, galvanizing public opinion against terrorism (Eid, 2014). Yet the continuous broadcast of these images plays into terrorist goals. Covering terrorism presents an ethical dilemma. A critical role for journalism is to keep the public informed about important issues (McCombs & Shaw, 1972), and citizens want to know about terrorist threats. Censoring coverage of such things does not play well against the norm of serving and informing the public, despite the concern over playing handmaiden to terrorist groups.

Under these conditions, we think it is important to reassess how major U.S. news outlets cover terrorist events. In particular, it is important to focus on which events get covered by major broadcast and cable television news outlets. We limit our focus to these outlets for two key reasons: (a) even today they have the widest audience reach among U.S. audiences, ostensibly making coverage by these select outlets especially attractive for publicity-seeking terrorist groups, and (b) television news has the properties most conducive for conveying the violence, drama, and carnage at the heart of terrorists' attacks and media tactics.

This study draws on a rich database of documented terrorist incidents prepared by the National Consortium for the Study of Terrorism and Responses to Terrorism (START), and a database of news content from six U.S. broadcast and cable networks: ABC, CBS, CNN, NBC, Fox, and MSNBC. Our data span 15 years, from 1998 to 2013. We model U.S. news coverage of terrorist attacks as a function of proximity to U.S. soil, affinity with the United States, the nature of attack, number of total and U.S. casualties, and the characteristics of the terrorist perpetrators. We find that the determinants of terrorism coverage are generally reflective of known norms and routines of journalism but that the relationships are more nuanced than expected.

TERRORISM AND MEDIA COVERAGE

Despite increases in terrorist activity since 2004, most instances of terrorism receive little or no media coverage (Campana, 2007; Chermak & Gruenewald, 2006). Of the 4 million words of domestic terrorism coverage published in the *New York Times* between 1980 and 2001, only 15 incidents accounted for 85% of the coverage (Chermak & Gruenewald, 2006). Campana's (2007) research on terrorism events between 1998 and 2005 shows that increased coverage across international newspapers follows only major global terrorism events. Terrorism coverage is primarily comprised of intense press reaction. The volume of "terrorism coverage" we see typically reflects intensive coverage of few events rather than moderate coverage of the high number of actual instances (Chermak &

Gruenewald, 2006; also see Powell, 2011). Chermak and Gruenewald (2006) offered four event characteristics that earn coverage: high casualties, links to domestic terrorist groups, targeting of airlines, and use of hijacking. Research on network evening news coverage of the TWA hostage crisis shows that the coverage surrounding the TWA hostage crisis was "dramatic, reactive, and extensive" (Atwater, 1987, p. 525). The seriousness of the event, generally measured in casualties, has a significant impact on coverage.

These findings comport with the norms and routines of journalism (Cook, 2005), the human negativity bias (Soroka, 2014), and the economics of news (Hamilton, 2004). Violence on television sells (Hamilton, 1998). Humans are predisposed to attend to negative and threatening information (Albertson & Gadarian, 2015; Soroka, 2014). Despite audiences' professed fatigue from negative, sensational news, countless studies verify the appeal of negative, sensational, and threatening information. Journalists make choices with these tendencies in mind (Gans, 1979).

> H1: Terrorist events with more casualties are likely to receive more U.S. media coverage.
> H2: Terrorist events with more U.S. casualties are likely to receive more U.S. media coverage.

Cultural relevance, affinity, and proximity also play key roles in determining newsworthiness (Gans, 1979). The same standards apply to terrorism. Domestic incidents are more likely to be covered than international (Chermak & Gruenewald, 2006; also see Jetter, 2015). International incidents are often judged to be less relevant to American viewers. Rohner and Frey (2007) and Pohl (2014) provided nuanced accounts of how terrorist media dynamics revolve around a balancing act between the level of risk and the level of media coverage. They asserted that, since 9/11, the media have devoted more attention to terrorism, which has encouraged an increase in terrorist activity. The media and the terrorist are engaged in a game in which they adjust their behavior according to the other. Their model predicts that terrorism attacks in North America and the West will attract significant media coverage regardless of the scale of attack. Given that our analysis focuses on U.S. media, this supports a proximity hypothesis:

> H3: Terrorist events in and proximate to the U.S are likely to receive more U.S. media coverage than terrorist events elsewhere.

News production is often affected by political and cultural factors (Gans, 1979; Shoemaker & Reese, 2011), such as the characteristics of the political systems in which the media outlets operate (e.g., Albæk, Van Dalen, Jebril, & de Vreese, 2014). News media's coverage of global terrorist events is influenced by the characteristics of the targeted nations in which an event occurs. Studies suggest a positive relationship between a nation's political affinity to the United States and the occurrence of terrorism events in that given nation (e.g., Dreher & Fischer, 2010). Using "voting coincidence" as an indicator for political affinity, Dreher and Gassebner (2008) found that, as U.S. voting coincidence increases, terror attacks are more likely for affiliated countries. Consistent with journalistic values of *geographic* proximity, attacks in *politically* proximate places are more likely to get coverage than elsewhere. News media's selection and presentation of terrorism events also depends on the "cultural proximity and cultural distance of the characteristics of a terrorist act" (Nossek & Berkowitz, 2006, p. 693). When the society's core values are threatened, journalists are more likely to emphasize the dominant cultural order; to the contrary, as the cultural distance increases, terrorism coverage becomes more symbolic (Nossek & Berkowitz, 2006). Because culturally proximate nations reflect core values and culture of the United States, attacks there are perceived as "our business," drawing interest and coverage from U.S. media. Covering what audiences consider "their business" is key to the market incentive structures of the U.S. media. We propose an *affinity* hypothesis:

H4: Terrorist events in countries with higher levels of affinity with U.S. are likely to receive more U.S. media coverage than events in countries with lower affinity with the U.S.

H5: Terrorist events in countries with which the U.S. regularly interacts are likely to receive more coverage by U.S. media than events in countries not well known.

Terrorism coverage also serves as an effective platform for policymakers and politicians (Chermak & Gruenewald, 2006). Powell (2011) found a pattern of coverage in which "fear of international terrorism is dominant, particularly as Muslims/Arabs/Islam working together in organized terrorist cells against a 'Christian America' while domestic terrorism is cast as a minor threat that occurs in isolated incidents by troubled individuals" (Powell, 2011, p. 91). This narrative was often repeated in the media by President George W. Bush, who frequently juxtaposed Iraq and 9/11 in his rhetoric, a theme repeated and developed in the media (Gershkoff & Kushner, 2005). Powell defined the resulting political climate as the "United States versus Islam that created animosity between East and West" and sustained a climate of fear of terrorism that is linked repeatedly to Muslims (Powell, 2011, p. 90). The climate of fear only feeds further speculation about the possibility of future attacks and generates increasing amounts of coverage across media networks (see Campana, 2007). Because our period of analysis is from 1998 to 2013, we expect that coverage of events is more likely when perpetrated by known groups associated with Islam that regularly challenge U.S. and Western interests.

H6: Terrorist events attributed to the Islamic State group/Taliban/Al-Quaida are likely to receive more coverage by U.S. media than events caused by other actors or groups.

Method of attack also determines whether terrorist events are newsworthy to journalists. Kydd and Walter's (2006) definition of terrorism as "the use of violence against non-state actors to attain political goals" (p. 52) suggests the importance of the "violence" characteristics to news media's coverage of terrorism. Violence almost guarantees media coverage and sensationalism (Hoffman, 2003). As the old adage of news production goes, "If it bleeds it leads." Crime and mass destruction are sensational topics (Hamilton, 1998; Uribe & Gunter, 2007), which attract wide public audiences despite the universal disdain (Davis & McLeod, 2003). Jetter's (2014) recent study shows that suicide attacks—especially suicide bombings—received significantly more media coverage than nonsuicide attacks. We offer the following hypothesis:

H7: Modes of attack that are capable of mass destruction, particularly gruesome, or novel are those most likely to earn coverage from major U.S. media outlets.

METHOD AND DATA

To examine how characteristics of terrorist events predict terrorism coverage, we combined the START Terrorism Database[1] with news coverage by six major U. S. networks—ABC, CBS, CNN, NBC, Fox, and MSNBC. START provides our predictors of interest, and coverage by news networks captures our dependent variables. The START data begin in the 1970s, but readily available news transcripts across the outlets of our analysis do not begin until 1998. Our data set spans 15 years, from 1998 to 2013; during this period, the total number of terrorism attacks increased annually and reached 11,952 in 2013, which is about 12.80 times the terrorist attacks in 1998 ($N = 933$).

START Terrorism Database

The START Terrorism Database is prepared by the National Consortium for the Study of Terrorism and Responses to Terrorism. It is an open-source database including statistical information on global terrorist incidents from 1970 to 2014. From 1998 to 2013, a total of 57,628 terrorist events occurred, resulting in 56,534 casualties and 54,829 injuries. These terrorism attacks happened in

---

[1] See online appendix A for more information about START.

more than 90 nations worldwide, concentrating in Iraq (20.75%), Pakistan (13.25%), and Afghanistan (10.12%), followed by India (9.26%), Thailand (4.22%), Philippines (4.03%), Russia (3.10%), and Colombia (2.98%).[2] Although these events were attributed to 1,275 perpetrator groups, the most "active" perpetrator was the Taliban (5.55%). The primary five types of attacks are bombing (54.42%), armed assault (25.12%), kidnapping/hostage (6.13%), assassination (6.10%), and facility attack (5.01%). Most attacks targeted on private citizens (26.58%), police (15.33%), and military (12.34%).

News Coverage Data

Using the START Terrorism Database (1998–2013) as our universe of incidents, we examine the extent to which terrorism events were covered by the major U.S. broadcast and cable news networks. We collected television news transcripts from ABC (17,841), CBS (26,753), NBC (18,487), CNN (96,478), Fox (24,942), and MSNBC (8,472).[3]

We developed a Python script to count the television news transcripts for each terrorism event. In addition, we asked four trained assistants to code a small sample of news transcripts—2018 in total from CBS and Fox in 2013—for the purpose of verification check, which suggests that the automatic coding is valid and conservative. See online appendix B for details about Python coding procedure and robustness checks using automatic and human-code data.

Variables and Measures

Our dependent variable (*terrorism coverage*) captures instances of U.S. news network coverage of each of the 57,628 terrorist events that occurred between 1998 and 2013. We counted the total number of news transcripts from six major cable and broadcast networks (range = 0–27, $M = 0.04$, $SD = 0.40$). Afterward, we computed a dichotomous indicator of coverage, with 1 representing "a terrorism event is at least covered by one network" (2.16%) and 0 representing "a terrorism event is covered by none of the six networks" (97.84%).

Our independent variables from the START data set relate to the characteristics of terrorist attacks. All variables are measured at the terrorist incident level. Measurement and descriptive statistics are located in online appendix C. The first predictor (*distance to U.S*) is measured in miles by calculating the distance

---

[2] The remaining countries where terrorism incidents occurred constituted less than 1% of the total attack locations.

[3] Examples of news programs and shows for analyzed news transcripts include *Nightline* (ABC), *NBC Nightly News* (NBC), *Today* (NBC), *CBS Evening News* (CBS), *Fox on the Record with Greta* (Fox), *The O'Reilly Factor* (Fox), *All In with Chris Hayes* (MSNBC), *Politics Nation* (MSNBC), *CNN Newsroom*, and *CNN Presents*.

between each event locale and Washington, DC, using latitude and longitude.[4] For all 57,628 terrorist events, distance ranged from 1.35 to 11569.03. We relied on Bailey, Strezhnev, and Voeten's (2015) data set to measure *(affinity to U.S.)*, which captures the affinity of the nation in which a terrorist event occurred to the United States, where −1 represents "least similar interest" and 1 represents "most similar interest" ($M = -0.30$, $SD = 0.23$). The larger the affinity score, the more policy preferences are shared.[5] *Global casualties* (range = 0–1381.5, $M = 2.37$, $SD = 11.36$) and *U.S. casualties* (range = 0–1357.5, $M = 0.07$, $SD = 7.98$) capture the total number of each resulting from terrorist events, respectively.

Dummy variables *(occurred in U.S. and other prominent nations)* capture whether a terrorist event occurred in the United States (0.60%), the United Kingdom (0.18%), India (9.26%), Afghanistan (10.12%), Iraq (20.75%), Libya (0.62%), Russia (3.10%), Syria (0.87%), and Pakistan (13.25%). These nations fit two criteria for testing our "regular interaction" hypothesis (H5): The United States regularly interacts with these nations, as friends or foes (Gans, 1979; Perlmutter & Hamilton, 2007; Wolfsfeld, 1997).

The measure for *primary perpetrator groups* captures whether an event is attributed to any of the three most prevailing perpetrator groups—Al-Qaida (23.04%), the Islamic State group (6.15%), and Taliban (54.26%). START gathers information on attribution of responsibility for attacks.[6] Our measure of *attack type* indicates the manner of attack: bombing (54.42%), hostage/kidnapping (6.13%), and hijacking (0.24%). *Suicide attack* indicates whether its a suicide attack (yes = 5.12%; no = 94.88%).

---

[4] Distance was calculated in Microsoft Excel using this formula: 6371*ACOS(COS (RADIANS(90-L1))*COS(RADIANS(90-L2))+SIN(RADIANS(90-L1))*SIN(RADIANS(90-L2)) *COS(RADIANS(G1-G2)))/1.609, where L1 refers to latitude and G1 refers to longitude of the city where an attack occurred, whereas L2 refers to latitude and G2 refers to longitude of Washington, DC. Geographic distance is 1.35 miles for terrorism attacks in Washington, DC, due to minor differences in DC latitude and longitude employed by START.

[5] The affinity score data by Bailey et al. (2015) was computed using two category vote data where 1 is approval for an issue and 2 is disapproval for an issue. Thus their affinity score represent political affinity rather than comprehensive affinity that is based on political, cultural, and other elements. Despite the absence of other dimensions, however, this affinity score itself still indicates certain difference between a nation and the United States.

[6] Readers might wonder whether START's reliance on print media means our dependent and independent variables are conflated. We do not think print coverage is determinant of the U.S. network and cable news coverage. The sheer volume of events in START that do not show up in our transcripts reflect that, as do the constraints of television news, which do not allow for the same depth and breadth of coverage afforded by print.

TABLE 1
Binary Logistic Regression Models Predicting Terrorism Coverage (1998–2013)

	Model 1 (Non-U.S. Cases Only)		Model 2 (All Cases)
	Baseline (1a)	Interaction (1b)	Baseline
No. of global casualties[a]	**0.51** (0.06)***	**0.47** (0.05)***	**0.34** (0.04)***
No. of U.S. casualties[a]	**0.70** (0.17)***	**0.71** (0.17)***	**0.72** (0.17)***
Distance to U.S.[a]	**−0.72** (0.12)***	**2.14** (0.98)*	—
Affinity to U.S.[a]	0.05 (0.13)	**−88.39** (28.16)**	—
Distance[a] × Affinity[a]	—	**10.27** (3.26)**	—
Occur in U.S.	—	—	**2.98** (0.35)***
Occur in India	—	—	−0.13 (0.28)
Occur in UK	—	—	**2.39** (0.54)***
Occur in Afghanistan	—	—	**0.96** (0.20)***
Occur in Iraq	—	—	**2.68** (0.12)***
Occur in Libya	—	—	**2.22** (0.36)***
Occur in Russia	—	—	**−2.01** (1.01)*
Occur in Syria	—	—	**1.37** (0.31)***
Occur in Pakistan	—	—	−0.12 (0.22)
Bombing	**3.03** (0.30)***	**3.08** (0.31)***	**2.68** (0.17)***
Hijacking	—	—	−1.41 (1.24)
Suicide	0.05 (0.18)	0.09 (0.17)	0.19 (0.10)
Al-Qaida	−0.04 (0.28)	−0.15 (0.28)	−0.21 (0.15)
The Islamic State group	**0.68** (0.11)***	**0.58** (0.11)***	−0.04 (0.18)
Taliban	**−1.66** (0.84)*	**−1.61** (0.81)*	**−0.81** (0.29)**
Constant	−0.14 (1.19)	**−24.52** (8.08)**	**−7.82** (0.19)***
N	38,004	38,004	52,241
Pseudo R^2	0.13	0.14	0.25

Note. Model 1 uses non-U.S. terrorism attacks, clustered by city of occurrence. Model 2 employs all cases. Entries are *coefficients* (robust standard errors). All independent variables are dummies, except for those with "a" superscript that are logged to curve for skewness.

[#]$p < .10$. *$p < .05$. **$p < .01$. ***$p < .001$. Drawn from two-tailed tests.

ANALYTICAL STRATEGY AND RESULTS

We focus on event characteristics that affect the probability of earning press coverage. Our dependent variable (*terrorism coverage*) is dichotomous, where 1 represents "earned coverage" and 0 otherwise. We utilize binary logistic regression, and results are shown in Table 1.[7] As expected, higher rates of global and U.S. casualities produce a higher likelihood of coverage. Translated into probability change, every one-unit increase in global causality boosts the probability of coverage up by about

---

[7] Rare event logistic regressions revealed same results; we report binary logistic models for better interpretation.

0.3%. Every one-unit increase in U.S. causality improves coverage probability by about 0.6%. These are significant and substantive changes, given that terrorism attacks often involve mass causalities. H1 and H2 are supported.

Consistent with H3, the more distant the event is from the United States, the less likely it is covered ($b = -0.72$, two-tailed $p < .001$). As distance from the United States increases across the range (from 351.04 miles to 10,241.5 miles),[8] the likelihood of coverage decreases by 7.30%. Model 2 also suggests the influence of cultural relevance to the United States on coverage. Relative to terrorist attacks occurring in the other nations left out of the model, those in the United States ($b = 2.98$, two-tailed $p < .001$), Britain ($b = 2.39$, two-tailed $p < .001$), Afghanistan ($b = 0.96$, two-tailed $p < .001$), Iraq ($b = 2.68$, two-tailed $p < .001$), Libya ($b = 2.22$, two-tailed $p < .001$), and Syria ($b = 1.37$, two-tailed $p < .001$) are more likely to get covered. There are exceptions: Russia ($b = -2.01$, one-tailed $p < .05$), India, and Pakistan. H5 is supported.

Attack type also matters: Weapons of mass destruction are more likely to earn coverage. A statistically significant and positive effect is found for "Bombing" (see baseline models in Table 1), revealing only partial support for H7. However, terrorist attacks initiated by prominent perpetrator groups (i.e., Al-Qaida) may not differ from attacks by the other groups. In fact, terrorism attacks initiated by Taliban are significantly less likely to be covered by U.S. national networks than those initiated by the perpetrators excluded from the model as the baseline. Terrorism attacks initiated by the Islamic State group are more likely to receive U.S. coverage ($b = 0.68$, one-tailed $p < .001$), but overall our findings are quite mixed, revealing only partial support for H6.

Recall that we expect an interaction effect between distance and political affinity, such that distant countries with a terrorism occurrence and high affinity to the United States are still likely to be covered: examples include Israel, Spain, Greece, England, and France as terrorism attacks occurring in these nations were covered by at least one of the six networks we chose for analysis. By contrast, terrorist attacks happening in nearby countries but with low political affinity to the United States—that is, Ecuador, Columbia, Mexico—were not covered by the networks.

Turning to Model 1b, we find a statistically significant two-way interaction effect between distance and affinity to the United States ($b = 10.27$, $p < .01$). Terrorist attacks occurring in nations distant from the United States are less likely to be covered than those happening in less distant nations, but the coverage probability for terrorism events in these distant nations is also conditional on their affinity with the United States, as shown in Figure 1.

---

[8] Note that this distance range is only for cases included in estimated model (baseline in Model 1, Table 1), such that it is different from the distance range for all cases, as discussed in the Measurement section.

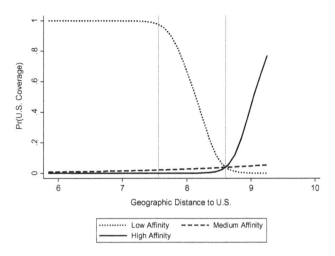

FIGURE 1 Predictive margins terrorism coverage probability, by distance and U.S. affinity.
Note. Figure is generated from the interaction model in Table 1. Distance and affinity are logged.

Overall, our analyses show strong support for the casualties hypotheses (H1/H2) and the proximity hypothesis (H3), as well as partial support for cultural relevance (H5), perpetrator (H6), and mode of attack (H7). Finally, our affinity hypotheses (H4) yielded mixed support. The coefficient for affinity in the baseline model was positive but failed to reach statistical significance. However, the interaction between affinity in distance in our second model is statistically significant and indicates that the effect of affinity is conditional. Also see online Appendix B for consistent results using alternative models and data.

DISCUSSION AND CONCLUSIONS

Like most forms of news, the decision to cover terrorist events is largely based on the interests and tastes of audiences, which—in the case of national news networks—is the entirety of the American mass public (see Robinson, 2007). The need to appeal to a mass audience means that certain characteristics of terrorist attacks predict whether major U.S. media cover them. Casualties, political and geographic proximity to the United States, violence, and destruction increase the likelihood of terrorism event coverage. Major U.S. news media are more likely to cover terrorism events in or proximate to the United States, when there are high rates of U.S. and global casualties, when major weapons are used, and when those with whom we are friendly are attacked. Given the market-based requirement to appeal to American audiences, the determinants of coverage observed are in line with expectations set forth by economic theories of news and the literature on the norms and routines of journalism (Cook, 2005; Hamilton, 2004).

The findings from our study also largely confirm previous studies on coverage of terrorism specifically. Chermak and Gruenewald (2006), Powell (2011), and Wittebols (1992) found that not all terrorist events are covered, despite the fact there has been an increase in terrorist events post 9/11. Our findings partially support the terrorism and media symbiosis hypothesis as explored in studies from Schmid and De Graaf (1982), Walsh (2010), Nacos (2007), Bockstette (2008), and Krueger (2008). Our results suggest that the U.S. media are more likely to cover non-U.S. events attributed to the Islamic State group, whose notoriety stems in part from their well-staged attacks. But we do not find similar effects for other well-known perpetrator groups. Still, there is evidence that groups' media sophistication (Pohl, 2014; Walsh, 2010), results in publicity. Our most consistent findings, however, support intuitions about what kinds of attacks are most likely to garner coverage and high volumes of it. It is clear from both the literature and our data that attacks on or proximate to U.S. soil are more likely to receive coverage and that the use of weapons of mass destruction and U.S. and global casualties predict the likelihood of coverage.

SUPPLEMENTAL DATA

Supplemental data for this article can be accessed here.

REFERENCES

Albæk, E., Van Dalen, A., Jebril, N., & de Vreese, C. H. (2014). *Political journalism in comparative perspective*. New York, NY: Cambridge University Press.

Albertson, B., & Gadarian, S. K. (2015). *Anxious politics: Democratic citizenship in a threatening world*. New York, NY: Cambridge University Press.

Atwater, T. (1987). Network evening news coverage of the TWA hostage crisis. *Journalism & Mass Communication Quarterly, 64*(2–3), 520–525.

Bailey, M. A., Strezhnev, A., & Voeten, E. (2015). Estimating dynamic state preferences from United Nations voting data. *Journal of Conflict Resolution, 17*, 1–27. doi:10.1177/0022002715595700

Bockstette, C. (2008). *Jihadist terrorist use of strategic communication management techniques*. Garmisch-Partenkirchen, Germany: George C. Marshall Center European Center for Security Studies. Retrieved from http://www.dtic.mil/cgi-bin/GetTRDoc?AD=ADA512956

Campana, P. (2007). Beyond 9/11: Terrorism and media in a mid-term period view (1998-2005). *Global Crime, 8*(4), 381–392. doi:10.1080/17440570701739744

Chermak, S. M., & Gruenewald, J. (2006). The media's coverage of domestic terrorism. *Justice Quarterly*, *23*(4), 428–461.

Cook, T. E. (2005). *Governing with the news: The news media as a political institution*. Chicago, IL: University of Chicago Press.

Davis, H., & McLeod, S. L. (2003). Why humans value sensational news: An evolutionary perspective. *Evolution and Human Behavior*, *24*(3), 208–216.

Dreher, A., & Fischer, J. A. (2010). Government decentralization as a disincentive for transnational terror? An empirical analysis. *International Economic Review*, *51*(4), 981–1002.

Dreher, A., & Gassebner, M. (2008). Does political proximity to the US cause terror? *Economics Letters*, *99*(1), 27–29.

Eid, M. (Ed.). (2014). *Exchanging terrorism oxygen for media airwaves: The age of terroredia: The age of terroredia*. IGI Global.

Gajanan, M. (2017, February 6). President Trump says media "doesn't want to report" on terrorist attacks. *Time*. Retrieved from http://time.com/4661625/president-trump-media-report-terrorism/.

Gans, H. J. (1979). *Deciding what's news: A study of* CBS Evening News, NBC Nightly News, Newsweek, and Time. New York, NY: Pantheon.

Gershkoff, A., & Kushner, S. (2005). Shaping public opinion: The 9/11-Iraq connection in the Bush administration's rhetoric. *Perspectives on Politics*, *3*(3), 525–537.

Hamilton, J. T. (1998). *Channeling violence: The economic market for violent television programming*. Princeton, NJ: Princeton University Press.

Hamilton, J. T. (2004). *All the news that's fit to sell: How the market transforms information into news*. Cambridge, MA: Harvard University Press.

Hoffman, B. (2003). *The logic of suicide terrorism*. Retrieved from http://www.theatlantic com/magazine/archive/2003/06/the-logic-of-suicide-terrorism/302739/.

Hoskins, A., & O'loughlin, B. (2007). *Television and terror: Conflicting times and the crisis of news discourse*. London, UK: Palgrave Macmillan.

Jetter, M. (2015). *Blowing things up: The effect of media attention on terrorism (No. 15-28)*. Retrieved from https://ecompapers.biz.uwa.edu.au/paper/PDF%20of%20Discussion% 20Papers/2015/DP%2015.28_Jetter1.pdf

Jetter, M. (2014). Terrorism and the media. IZA Working Paper #8497.

Krueger, A. B. (2008). *What makes a terrorist: Economics and the roots of terrorism*. Princeton, NJ: Princeton University Press.

Kydd, A., & Walter, B. (2006). The strategies of terrorism. *International Security*, *31*(1), 49–80.

McCombs, M. E., & Shaw, D. L. (1972). The agenda-setting function of mass media. *Public Opinion Quarterly*, *36*(2), 176–187.

Mitchell, A., Gottfried, J., Barthel, M., & Shearer, A. (2016, July 7). *The modern news consumer, Pathways to news*. Pew Research Center. Retrieved from http://www.journalism.org/2016/07/07/pathways-to-news/.

Nacos, B. L. (2007). *Mass-mediated terrorism: The central role of the media in terrorism and counterterrorism*. Lanham, MD: Rowman & Littlefield.

Nossek, H., & Berkowitz, D. (2006). Telling "our" story through news of terrorism: Mythical newswork as journalistic practice in crisis. *Journalism Studies*, *7*(5), 691–707.

Perlmutter, D., & Hamilton, J. (2007). *From pigeons to news portals: Foreign reporting and the challenge of new technology*. Baton Rouge, LA: Louisiana State University Press.

Pohl, G. (2014). *A new picture for understanding terrorists' opportunities and choices when media coverage is a desired payoff*. Retrieved from https://papers.ssrn.com/sol3/papers2 cfm?abstract_id= 2411322.

Powell, K. A. (2011). Framing Islam: An analysis of U.S. media coverage of terrorism since 9/11. *Communication Studies*, *62*(1), 90–112. doi:10.1080/10510974.2011.533599

Robinson, M. J. (2007, August 22). *Two decades of American news preferences*. Retrieved from http://www.pewresearch.org/2007/08/22/two-decades-of-american-news-preferences-2/

Rohner, D., & Frey, B. S. (2007). Blood and ink! The common-interest-game between terrorists and the media. *Public Choice, 133*(1–2), 129–145.

Schmid, A. P., & De Graaf, J. (1982). *Violence as communication: Insurgent terrorism and the Western news media*. London, UK: Sage.

Shoemaker, P., & Reese, S. D. (2011). *Mediating the message*. New York, NY: Routledge.

Soroka, S. N. (2014). *Negativity in Democratic politics: Causes and consequences*. New York, NY: Cambridge University Press.

Uribe, R., & Gunter, B. (2007). Are sensational news stories more likely to trigger viewers' emotions than non-sensational news stories? A content analysis of British TV news. *European Journal of Communication, 22*(2), 207–228.

Walsh, J. I. (2010). *Media attention to terrorist attacks: Causes and consequences*. Retrieved from https://pdfs.semanticscholar.org/8c8a/80c619cca15d5b5848ad762fd8ec9ab962bf.pdf.

Wittebols, J. H. (1992). Media and the institutional perspective: U.S. and Canadian coverage of terrorism. *Political Communication, 9*(4), 267–278.

Wolfsfeld, G. (1997). *Media and political conflict: News from the Middle East*. New York, NY: Cambridge University Press.

APPENDIX A. START AND ITS DEFINITION OF TERRORISM EVENTS

This study employs the global terrorism database prepared by the National Consortium for the Study of Terrorism and Responses to Terrorism (START) as the avenue for our independent variables as well as the universe for capturing terrorism coverage.

The START terrorism database is "an open-source database including information on terrorist events around the world from 1970 through 2015" (START, 2016). It defines a terrorist attack as "the threatened or actual use of illegal force and violence by a non-state actor to attain a political, economic, religious, or social goal through fear, coercion, or intimidation" (Global Terrorism Database, 2016, p. 9). The three essential criteria START uses to define a terrorism event include: intentional, violent, and non-state based.[1]

For the past 45 years, the START has documented over 150,000 global terrorist events. For each of them, START records a set of information including the date and location of occurrence, the weapons used, the nature of attack target/victim, the perpetrators of terrorism incidents,[2] and the number of casualties and wounds involved in each case,[3] etc. Specifically, at least 45 variables are recorded for each terrorism incident, and for more recent incidents over 120 variables are captured.

START draws such information from a variety of open and credible media sources, using a combination of automated and manual data collection strategies. Specifically, the START database uses "machine learning and data mining techniques to identify and coding the incidents that are included in the GTD; using machine learning and data mining techniques to identify news articles that include information about terrorist attacks; and developing and utilizing a proprietary Data Management System (DMS) to compile the database" (Global Terrorism Database, 2016). Its data collection process "begins with a universe of over one million media articles on any topic published daily worldwide in order to identify the relatively small subset of articles that describe terrorist attacks" (Global Terrorism Database, 2016, p. 7). In addition to English-language news, START also takes advantage of English-language news stories translated from news of over 80 languages published in over 160 countries.

---

[1] See GTD Codebook for more criteria START employs to define a terrorist event.

[2] To ensure consistent usage of group names, START established standardized list (Global Terrorism Database, 2016). While START records multiple perpetrators in cases where responsibility for the attack is attributed to more than one actor or group, we focus on the primary perpetrator.

[3] START documents all victims and attackers who died as a direct result of the incident. It also employs multiple approaches to maximize statistical accuracy: first, it relies on multiple independent sources' most up-to-date reports to draw the number of casualties; second, the majority figure is used when multiple sources published around the same time or the validity of any single source is questionable; third, casualty numbers on claims made by a perpetrator group are not used; fourth, a casualty figure is not recorded if it comes from a source of questionable validity; lastly, these two variables were coded as "missing" if the figure of fatalities is not reported or if it is too vague for use (Global Terrorism Database, 2016).

START employs a variety of approaches to ensure the validity of the terrorism data. For example, START draws terrorism information from high-quality sources, which are often "independent (free of influence from the government, political perpetrators, or corporations) ... routinely report externally verifiable content" (Global Terrorism Database, 2016, p. 7). If several sources document the same terrorism incident, at least one of them must be high-quality source. In addition, terrorism events are not included in the START database if they are *only* documented by distinctly biased or unreliable sources. Also due to this appeal of using high-quality sources, the terrorism data documented by START are quite conservative that may not reflect all incidents. In addition, to improve the validity of statistical information, START also updates the database with new documentation about an event, as necessary and appropriate.

Check the START webpage (http://www.start.umd.edu/gtd/about/) and GTD codebook (2016) for more details and information.

APPENDIX B. PYTHON CODING PROCEDURE AND ROBUSTNESS CHECKS

Scholars more regularly use computational methods together with manual methods in large scale content analysis studies (see Lewis, Zamith & Hermida, 2013 for a full review). In order to match a substantial number of news articles with a total of 57,628 terrorist events that occurred from 1998 to 2013, we opted to employ a computational approach in this study. The procedure of automatic coding is described below:

As a starting point, we created a spreadsheet to organize the terrorism events by date (day, month, and year) and event ID.[4] For each terrorism event, another four sets of information were also included: the city and country of occurrence, the weapon type and attack type as recorded by START. As we'll also discuss later, these four criteria were used to assess whether a news transcript covers each terrorism event. Once the terrorism events data were reorganized, we resaved them into a CSV file for further analysis.

Our news transcripts for the six selected U.S. networks were drawn from Lexis-Nexis[5] and saved in TXT format. Each text file was renamed to represent the news source and the year from which the news transcripts were drawn, i.e. "ABC_1998." Note that a number of news transcripts were included in each text file, such that for automatic coding we used "All rights reserved" as a separator

---

[4] Event ID is originally assigned by START, which suggests the sequence of terrorism attacks by day, month, and year.

[5] Transcripts for these news networks were accessed through Lexis-Nexis using the search term "terror." With the intention to capture each network's overall coverage of terrorism attacks, we didn't purposively focus on any specific news programs; instead, we drew transcripts from all news programs that are documented by Lexis-Nexis. As Table 1 (see main manuscript) displays, the news programs from which news transcripts were drawn are mostly major news programs, with more news shows for cable networks than for national networks.

to signal the end of one news story as well as the start of next new story. We chose "All rights reserved" because it was consistently used at the very end of each news transcript, which serves as a most valid and reliable separator to distinguish one news transcript from another.

After these two steps, we ended up two sets of information: one set of terrorism events (saved as CSV), and another set of news transcripts (saved as TXT). We then composed a Python script to match the news transcript data in accordance with the terrorism event data. We selected Python because it is a powerful programming language for parsing large-scale data (Lewis, Zamith & Hermida, 2013). As all news transcripts were reorganized by year and source i.e. "ABC_1998," Python first relied on the name of TXT files to decide which transcripts should be used to account for whether a terrorism event was covered by a certain network – ABC, CBS, CNN, FOX, MSNBC, and NBC, respectively. Afterward, using "All rights reserved" as a separator, Python automatically divides each TXT file into a set of news transcripts with metadata information including date and publisher. For each single news transcript, our Python script followed a fairly conservative rule to minimize the error margin of automatic coding (false positives in particular):

a. First, the match of terrorism events with terrorism coverage was limited to each corresponding calendar year – January 1st to December 31th for 1998 through to 2013, respectively – such that we did not take into account news coverage in following years;
b. Afterward, for each terrorism event, we only counted the number of news transcripts broadcast in the immediate 7 days following its occurrence (including the occurrence day). For example, if a terrorism event occurred on January 1, 1999, we counted how many transcripts were broadcast from January 1, 1999 to January 7, 1999 covered it;
c. Next, we utilized two criteria – location of terrorism events and the manner of attack – to match terrorism events with news transcripts. Specifically, a news transcript was coded as a terrorism coverage piece only if it mentioned (a) the specific city and (b) nation in which a terrorism event occurred, as well as (c) the attack type used in the event.[6]

---

[6] Note that we also matched the START incidents with news transcripts using "weapon type" to replace "attack type" while keeping country and city criteria the same. We assumed that the weapon type – which is often more specific and diverse – would better capture terrorism events. However, eyeball check shows that certain weapon categories used by START are not commonly used in the news. For example, for the 9/11 terrorist attack, the weapon type was recorded as "melee" in START, which was not used by any networks at all. This not only causes the failure to capture the 9/11 terror event but also indicates that the "weapon type" criterion may be restrictive in capturing news transcripts. Further comparison shows that the data drawn with "weapon type" do slightly differ from the data drawn with "attack type," as shown in robustness check 3 below. These slight variations may indicate that despite the different criteria, Python scripts work quite well in matching the news articles with terrorism events.

ROBUSTNESS CHECK 1: PYTHON CODING AGAINST HUMAN CODING

To check the validity of automatic coding by Python, we conducted two sets of robustness checks by comparing Python coding with human coding.

Descriptive Statistics

First, we looked at the descriptive statistics of Python coding and human coding. Results show that much fewer news transcripts were coded as covering terrorism event by Python than by humans, suggesting that our automatic coding is more conservative. This is also supported by the number of terrorism coverage pieces coded by Python: as Figure 1 demonstrates, despite a large number of television news transcripts drawn from Lexis-Nexis, only a small number of them were coded as covering terrorism events, with a minimum of 0 (in 2000) and maximum of 616 (in 2007). Moreover, when checking the automatic coding by networks, it is consistent with our expectation that CNN was found to have the largest volume of terrorism coverage (2,057), followed by CBS (79), FOX (71), MSNBC (69), ABC (43), and NBC (40). Specifically, a total of 13 domestic cases were covered by U.S. networks, such as 9/11 and Boston marathon terror attacks (see more details in online appendix D).

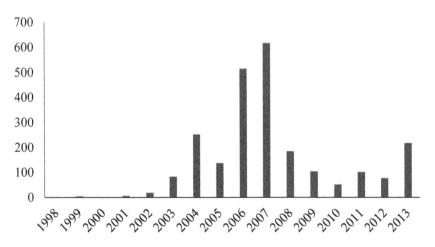

FIGURE 1 Sum of Television News Transcripts that Covered Terrorism Events, 1998 to 2013

Source: ABC, CBS, CNN, NBC, FOX, and MSNBC. Note: Number of television news transcripts by year is based on the sum of stories from all six networks.

Regression Models

To further check for the validity of automatic coding by Python, we also conducted four sets of truncated logistic regression models[7] with two small-scale datasets where one documents the news coverage outcome variable using human coding while the other documents automatic coding.

Data and Procedure:

For both datasets, the universe of terrorism events are 11,925 incidents drawn from START's 2013 subset. News transcripts were drawn from CBS and FOX's 2013 archive, which were searched in LexisNexis using "terrorism" as the search word. Afterward, we followed the Python procedure as described earlier to generate the automatic coding of the number of news transcripts covering each incident; on the other hand, four student coders matched over 2,000 news transcripts with the 2013 terrorism events. Thus in both datasets, the total number of observations (N = 11952) and the independent variables are exactly the same, with the only variance in the number of news transcripts covering each single terrorism event (Human coding: range = 0–237, M = 0.03, SD = 2.23; Automatic coding: range = 0–1, M = 0.002, SD = 0.05).

As the outcome variable "Terrorism Coverage" is dichotomous, we conducted binary logistic regression models predicting the likelihood that a terrorism event is covered by U.S. media. While we tried to replicate the models as reported in the main manuscript, different variables were dropped automatically in both models due to the fact that this given variables predicts failure perfectly. This gives rise to the concern that such models are not comparable. Instead, to allow for direct comparison of the coefficient estimates generated by human-versus automatic-coding data, we conducted four sets of equivalent models to as displayed in Table B-1. These truncated models focus on our most important predictors – "distance to U.S." and the interaction effect between geographical distance and political affinity. Again, as our models include "affinity score", we focused our analysis on non-U.S. cases only.

Findings:

The results reveal consistent patterns of our most important predictors – "distance to U.S." and the interaction effect between geographical distance and political affinity. As demonstrated in Table B-1, "distance to U.S." was found to have statistically significant negative effect on the probability of U.S. terrorism

---

[7] Note that the START data have some missing values. For example, for the predictors we're interested in, the rate of missing values is 31.68% for "geographic distance," 14.5% for "political affinity," 12% for "global casualty" and around 10% for the other included predictors.

TABLE B-1

Truncated Equivalent Models Predicting Likelihood of Terrorism Coverage, Human vs. Python Coding (Source: CBS and FOX; 2013)

	Human Coding		Python Coding	
	Baseline	Interaction	Baseline	Interaction
Distance to U.S.[a]	**−1.90****	0.98	**−1.18***	1.82
	(0.66)	(1.90)	(0.57)	(1.47)
Affinity to U.S.[a]	−1.33	−101.14*	−1.03	−94.74*
	(1.68)	(51.17)	(0.84)	(38.20)
Distance[a] **X Affinity**[a]	—	**11.55***	—	**10.81***
		(5.90)		(4.45)
Constant	10.21	−14.36	3.72	−21.97
	(5.33)	(16.34)	(4.93)	(12.31)
N	11,722	11,722	11,721	
Pseudo R^2	0.02	0.03	0.01	

Note: All models use non-U.S. terrorism attacks only, which are clustered by city of occurrence. Entries are *coefficient* (robust standard error). All independent variables are dummies, except for those with "a" superscript that are logged to curve for skewness. $*p < 0.05$, $**p < 0.01$, and $***p < 0.001$ are drawn from two-tailed tests.

coverage (Human Coding: b = −2.17, $p < 0.05$; Automatic Coding: b = −1.52, $p < 0.05$). In addition, the coefficient on the "distance X affinity" interaction term is statistically significant and suggest the same directions (Human Coding: b = 11.55, $p < 0.05$; Automatic Coding: b = 10.81, $p < 0.05$).

As such, despite differences in the number of news transcripts covering the 11,952 terrorism events occurring in 2013, the models conducted with automatic coding and human coding datasets still reveal quite consistent results, which not only provides additional evidence for the validity of Python coding but also suggests the robustness of our most intriguing finding regarding the "distance X affinity" interaction effect.

ROBUSTNESS CHECK 2: DATA FROM "ATTACK TYPE" VARIANT OF PYTHON CODE, TWO-DAY CYCLE

While the above comparison suggests the validity of automatic coding, we seek to provide more evidence for the robustness of our findings by using more conservative automatic-coding data. To that end, we followed the same coding procedure as described above; however, instead of matching news transcripts with terrorism attacks occurring within the prior one week, we restricted automatic match to two days –news transcripts' publishing day and the prior day. This greatly reduced the number of terrorism events covered by U.S. networks: relative to 2.16% of the 1998–2013 terrorist attacks that received U.S. news coverage when coded within one-week cycle, only 0.87% were covered when using two-day cycle.

We then replicated the same models as reported in main manuscript, with the results displayed in Appendix Table B-2. When compared with the results reported in main manuscript Table 2, one can find a high consistency in terms of both the direction and statistical significance of almost all coefficients. Also, the magnitude of most coefficients did not vary much either. This provides more support for our robust findings.

TABLE B-2

Binary Logistic Regression Models Predicting Terrorism Coverage (1998–2013), using More Conservative Automatic Coding data[a]

	Model 1 (Non-U.S. Cases Only)		Model 2 (All Cases)
	Baseline	Interaction	Baseline
# of Global Casualties[a]	**0.54**(0.04)***	**0.51**(0.04)***	**0.40**(0.05)***
# of U.S. Casualties[a]	**0.88**(0.19)***	**0.89**(0.19)***	**0.89**(0.20)***
Distance to U.S.[a]	**−0.61**(0.12)***	**2.30**(0.90)*	—
Affinity to U.S.[a]	0.13(0.14)	**−87.64**(25.34)**	—
Distance[a] X Affinity[a]	—	**10.19**(2.92)***	—
Occur in U.S.	—	—	**3.15**(0.41)***
Occur in India	—	—	0.48(0.30)
Occur in UK	—	—	—
Occur in Afghan	—	—	**0.68**(0.29)*
Occur in Iraq	—	—	**2.34**(0.17)***
Occur in Libya	—	—	**1.74**(0.60)***
Occur in Russia	—	—	−1.41(1.01)
Occur in Syria	—	—	**1.07**(0.45)*
Occur in Pakistan	—	—	−0.28(0.31)
Bombing	**3.14**(0.20)***	**3.19**(0.20)***	**2.90**(0.28)***
Hostage	—	—	—
Hijacking	—	—	—
Suicide	0.09(0.09)	0.14(0.09)	0.23(0.14)
Al-Qaida	−0.42(0.34)	−0.52(0.34)	**−0.56**(0.24)*
ISIL	**1.56**(0.10)***	**1.47**(0.10)***	**0.86**(0.19)***
Taliban	**−2.97**(0.91)**	**−2.92**(0.89)**	**−2.14**(0.75)**
Constant	−2.15(1.39)	**−26.98**(7.30)***	**−8.70**(0.29)***
N	38,319	38,319	52,847
Pseudo R^2	0.13	0.14	0.22

Note: Model 1 uses non-U.S. terrorism attacks, which is clustered by city of occurrence. Model 2 employs all cases including U.S. attacks. For both, entries are *coefficient* (robust standard error). All independent variables are dummies, except for those with "a" superscript that are logged to curve for skewness. *$p < 0.05$, **$p < 0.01$, and ***$p < 0.001$ are drawn from two-tailed tests.

---

[a] Note that our models also include "Hostage" as a predictor, which was automatically dropped (3114 obs not used) as it predicts failure perfectly.

ROBUSTNESS CHECK 3: DATA FROM "WEAPON TYPE" VARIANT OF PYTHON CODE, ONE-WEEK CYCLE

Data for analysis here were drawn using the same coding procedure as described above; however, we employed "weapon type" instead of "attack type" as a matching criterion. Again, the results (see appendix Table B-4) were highly consistent.

TABLE B-3
Terrorist Incidents and Terrorism Coverage from 1998 to 2013, Using Alternative Data Drawn with "Weapon Type"

	Descriptive Statistics	*Source*
Count of Terrorism Coverage	Range: 0–19	Lexis-Nexis transcripts
	M = 0.04, SD = 0.37	ABC, CBS, NBC, CNN,
Terrorism Coverage	1 = Yes (1.90%)	MSNBC and FOX
	0 = No (98.10%)	
Distance to U.S. (in miles)	Range: 1.35-11569.03	
	M = 6403.46, SD = 1533.80	
Affinity to U.S	Range: −0.65–1.00	
	M = −0.30, SD = 0.23	
Nations of Attacks		
United States	0.60%	
United Kingdom	0.18%	
India	9.26%	
Afghan	10.12%	
Iraq	20.75%	
Libya	0.62%	
Russia	3.10%	
Syria	0.87%	
Pakistan	13.25%	
Perpetrator Groups		START Terrorism Database
Al-Qaida	23.04%	
Taliban	54.26%	
ISIL	6.15%	
Primary Attack Types		
Bombing	54.42%	
Hostage/Kidnapping	6.13%	
Hijacking	0.24%	
Suicide Attacks	Yes = 5.12%	
	No = 94.88%	
Terrorism Casualties		
# of Global Casualties	Range: 0-1381.5	
	M = 2.37, SD = 11.36	
# of U.S. Casualties	Range: 0-1357.5	
	M = 0.07, SD = 7.98	

Note: All variables are measured on terrorist incident level.

TABLE B-4

Binary Logistic Regression Models Predicting Terrorism Coverage (1998–2013), Using Alternative Data Drawn with "Weapon Type"[9]

	Model 1 (Non-U.S. Cases Only)		Model 2 (All Cases)
	Baseline	*Interaction*	*Baseline*
# of Global Casualties[a]	**0.48**(0.06)***	**0.44**(0.06)***	**0.31**(0.05)***
# of U.S. Casualties[a]	0.56(0.33)#	0.58(0.33)#	**0.53**(0.23)*
Distance to U.S.[a]	**–0.75**(0.11)***	**2.16**(0.99)*	—
Affinity to U.S.[a]	0.09(0.16)	**–89.40**(26.56)**	—
Distance[a] X Affinity[a]	—	**10.39**(3.06)**	—
Occur in U.S.	—	—	**3.62**(0.60)***
Occur in India	—	—	**–2.63**(0.99)**
Occur in UK	—	—	**3.16**(0.61)***
Occur in Afghan	—	—	**0.77**(0.36)*
Occur in Iraq	—	—	**3.00**(0.34)***
Occur in Libya	—	—	**2.41**(0.38)***
Occur in Russia	—	—	–0.29(0.69)
Occur in Syria	—	—	1.28(0.80)
Occur in Pakistan	—	—	–0.15(0.42)
Bombing	**1.61**(0.14)***	**1.66**(0.15)***	**1.17**(0.27)***
Hostage	0.34(0.18)#	0.42(0.17)*	0.34(0.28)
Hijacking[10]	—	—	–3.56(3.31)
Suicide	0.27(0.16)#	0.32(0.15)*	0.36(0.28)
Al-Qaida	0.15(0.46)	0.04(0.46)	–0.05(0.25)
ISIL	**–2.21**(0.17)***	**–2.30**(0.17)***	**–2.94**(0.37)***
Taliban	**–1.96**(0.84)*	**–1.91**(0.81)*	–0.71(0.44)
Constant	1.29(1.12)	–23.49(8.10)**	-6.82(0.30)***
N	39,641	39,641	55,357
Pseudo R[2]	0.09	0.10	0.24

Note: Model 1 uses non-U.S. terrorism attacks, which is clustered by city of occurrence. Model 2 employs all cases including U.S. attacks. For both, entries are coefficient (robust standard error). All independent variables are dummies, except for those with "a" superscript that are logged to curve for skewness. #$p < 0.10$, *$p < 0.05$, **$p < 0.01$, and ***$p < 0.001$ are drawn from two-tailed tests.

---

[9] Though not reported here, due to the failure to capture the 9/11 terrorism event, we conducted exactly the same models excluding the two observations for 9/11, which revealed similar results as reported here. Also, we conducted models controlling for the target of attacks (i.e. private citizens, police, military, or education), and they also revealed the same results.

[10] Note that our models also include "Hijacking" as a predictor, which was automatically dropped (45 obs not used) as it predicts failure perfectly.

APPENDIX C. DESCRIPTIVE STATISTICS

TABLE C-1
Terrorist Incidents and Terrorism Coverage from 1998 to 2013

	Descriptive Statistics	*Source*
Count of Terrorism Coverage	Range: 0–27	Lexis-Nexis transcripts for
	M = 0.04, SD = 0.40	ABC, CBS, NBC, CNN,
Terrorism Coverage	1 = Yes (2.16%)	MSNBC and FOX
	0 = No (97.84%)	
	Range: 1.35-11569.03	
Distance to U.S. (in miles)	M = 6403.46, SD = 1533.80	
Affinity to U.S	Range: –0.65–1.00	
	M = -0.30, SD = 0.23	
Nations of Attacks[11]		
United States	0.60%	
United Kingdom	0.18%	
India	9.26%	
Afghan	10.12%	
Iraq	20.75%	
Libya	0.62%	
Russia	3.10%	
Syria	0.87%v	
Pakistan	13.25%	
Perpetrator Groups		START Terrorism Database
Al-Qaida	23.04%	
Taliban	54.26%	
ISIL	6.15%	
Primary Attack Types		
Bombing	54.42%	
Hostage/Kidnapping	6.13%	
Hijacking	0.24%	
Suicide Attacks	Yes = 5.12%	
	No = 94.88%	
Terrorism Casualties		
# of Global Casualties	Range: 0-1381.5	
	M = 2.37, SD = 11.36	
# of U.S. Casualties	Range: 0-1357.5	
	M = 0.07, SD = 7.98	

Note: All variables are measured on terrorist incident level.

---

[11] In this study we defined culturally relevant nations as those with whom the U.S. regularly interacts and is therefore well known to U.S. audiences. While we tried to capture as many prominent nations that frequently interact with the U.S. as possible, the START data are limited in many ways that narrowed our focus to these nations included in our models. For example, while China frequently interacts with the U.S., only 35 out of 63,890 terrorism events occurred there from 1998 to 2013, which counts for only 0.05% of the total attacks.

Index

185